Also in the Pira guide series

Total quality management
Colour reproduction in the printing industry
Management information systems for printers
Wastepaper
Colour control for lithography
Printing inks – formulation principles and quality control testing procedures

Pira Guide Series

Print Management

Derek Porter
Principal lecturer, School of Printing Technology, London College of Printing and Distributive Trades

Pira International

Hardback reprinted 1995

ISBN 1 85802 021 2 (hardback)
ISBN 0 07 709001 2 (paperback)

First published by
Pira International
Randalls Road
Leatherhead
Surrey
KT22 7RU
UK

Paperback published in association with McGraw-Hill Book Company Europe, Shoppenhangers Road, Maidenhead, Berkshire SL6 2QL

Typeset in the UK by Manton Typesetters,
5-7 Eastfield Road, Louth, Lincolnshire.
Printed in the UK by The Eastern Press Limited,
14 Portman Road, Reading, Berkshire.

Contents

Preface

This book has been written for all those who have an interest in print management. It therefore concentrates on the key disciplines involved in successful print management. Each chapter introduces a discipline of which a manager, or intending manager, must have a working knowledge.

Dramatic advances are continually being made in printing technology and these present a major challenge to all managers. Print companies are faced with demands for increasingly high quality and intense price competition. There is thus a need for new management skills and new management attitudes towards technology and innovation.

Many print managers have, in the past, tended to concentrate on a particular area, such as production or sales, thus acquiring little knowledge about the overall management function. However, it is only through an integrated knowledge of management functions and techniques that a manager can realize his/her own potential, and his/her company's potential.

This guide is designed to provide a clear understanding of the responsibilities and nature of modern print management, where 'print' includes all the aspects and processes of the printing industry.

1

Printing-company organization: the role and scope of the printing manager

PRINTING MANAGEMENT

It is a fact of printing industry life that, while rapidly advancing technology in all fields of print production continues, human skills and attitudes, and the control of these skills and attitudes, are becoming more important than ever before.

To assess the skills involved in printing management it is first necessary to examine precisely what is to be managed. If we accept that management should take place within an organizational structure, it becomes logical that attention must first be given to establishing the nature of printing as an industry, and the nature of a company engaged in various printing processes.

The printing industry is a highly competitive and demand-driven business, and it currently ranks as the sixth largest industry in the UK. Approximately 170 000 people are employed in 10 000 companies spread throughout the UK.

It is difficult to identify any part of the national economy that is not served by the printing industry. Managing a printing company in a changing technological environment, often against intense local, national, or international competition, is a task that calls for considerable skill and judgment. Therefore, we need to ask ourselves what exactly our company is. What are its functions and how does it currently operate? Finally, how can it be controlled to allow progression? Only when these questions are properly answered will it be possible to concentrate on examining and assessing operating and other problems and the possibilities that exist to improve the management of the situation.

MANAGEMENT PRINCIPLES

In contrast to the situation in other sciences and activities, printing managers have no access to recognized standards, as the appropriate standards vary with the situation. Therefore, we can view management standards as guide statements.

Henry Ford stated the ten truths of management as follows:

- Think before you act; it is not always your money.
- All good management is the expression of one good idea.
- No good manager devotes effort to proving him/herself wrong.
- Cash in must exceed cash out.
- Management capability is always less than the organization actually requires.
- If complicated calculations are needed to justify an action, do not do it.
- An employee or manager either can do the job, or cannot do the job.
- If something is done wrong, it will be done badly.
- If you are attempting the impossible, you will fail.
- The easiest way to make money is to stop losing it.

MANAGEMENT FUNCTIONS

Henri Fayol, a French mining engineer and researcher in management, analysed management activity in terms of five elements[1]:

- planning,
- organizing,
- commanding,
- coordinating,
- controlling.

Further, for these elements to be effectively carried out, it is necessary to divide the elements into 14 principles as follows:

- authority that must match responsibility,
- discipline,
- the division of work and specifications,
- unity of command,
- unity of direction,
- remuneration that must be fair in relation to effort,
- centralization,
- the hierarchical principle of line authority,
- the subordination of individual interests to the general interest,
- equity,
- the principle of order,
- stability of tenure for the staff,
- the importance of initiative,
- the importance of belonging to a team.

ORGANIZATIONAL CRITERIA

Obviously, some organizational criteria are common to all types of business, regardless of the type of work the company is engaged in. Even so, effective management of these common criteria is critical to production through effective organization. Peter Drucker[2] states that:

> A manager organizes. He or she analyses the activities, decisions, and relations needed. He classifies the work, he divides it into manageable activities and further divides the activities into manageable jobs. He groups these units and jobs into an organizational structure. He or she selects people for the management of these units and for the jobs to be done.

The first common criteria are the purpose and aims of the business. For a printing company, these consist in providing the client with printed or other material within the time required, in the right place, at the correct quality and quantity, at the right price, and finally, most importantly, at an acceptable profit to the company. Considerable emphasis is placed upon profit, for the generation of profit is seen as the ultimate objective of printing-company business. An emphasis on profit alone does, however, tend to create a rather over-simplistic view of the process of printing-company management, as no general hierarchy of functional importance applies to all companies. Therefore, it is necessary to consider the following:

- the corporate strategy of the company, in which the aims and objectives relating to the future operations of the company are defined as policy; the aims and objectives must take into account legal and social structures and marketing proposals,
- the technology already available for production, or that planned for installation within the near future, as this will have a direct influence on:
    - the capital required,
    - the production capacity, both that existing and that created through planned development,
    - the production organization and methods,
    - the implications of the production workers' skills and attitudes to productivity.

Another common characteristic of printing organizations relates to people. No company can operate without people acting as the functional staff. No organizational aims or objectives can be achieved without people. An assessment of the skills involved in print production can be carried out via job evaluation and analysis (see Chapter 4: Human resources). A further consideration is the technology and equipment that is available within the printing company to achieve production. The management of production, and therefore productivity, is also concerned with the technology of processes. It must be organized according to the technology available. It is here that a major problem can arise, as printing technology has made tremendous advances in the last few years, and continues to do so. As a consequence, the capital cost of equipment and machinery has increased enormously, making it imperative to obtain the maximum effective utilization of machinery. Therefore, some definite strategies must be developed in order to achieve good overall company performance.

A final characteristic is the organization structure and the management style. People differ in attitude, personality, and ambition; some are leaders, and others need to be led. Achieving organizational objectives through the efforts of staff results in some people attaining the authority to supervise others.

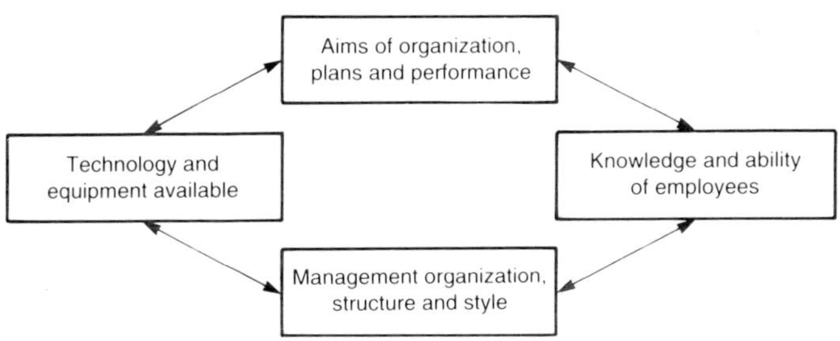

Figure 1.1 Management organization model

This means exercising the skills and responsibility for planning, controlling and directing the activities of staff through an organizational structure.

The style of management adopted is crucial to achieving the company's aims through the coordination of client requirements, the staff skills available, and the standard of technology to be used.

The situation is shown in Figure 1.1.

SKILLS REQUIREMENTS

The skills and functions required for efficient printing management are as follows:

- planning and organizational skills that achieve company policy (these combine the agencies of land, property, technology, labour and capital),
- the skills of understanding and developing company strategy and policies in the short and long term,
- initiative, innovation and judgment skills,
- technical knowledge relating to the production methods, processes and technology used,
- system-thinking skills entailing adaptation to and implementation of the management system,
- production-planning and management skills entailing the continuous checking of performance, and the taking of any necessary action,
- interpersonal skills, i.e. the ability to work with and control others, operating in conjunction with communications skills,
- environmental awareness and perception,
- behaviour-analysis skills that facilitate effective human-resources control that furthers the company's aims and objectives,
- market research and analysis skills,
- information collection, evaluation and analysis skills,

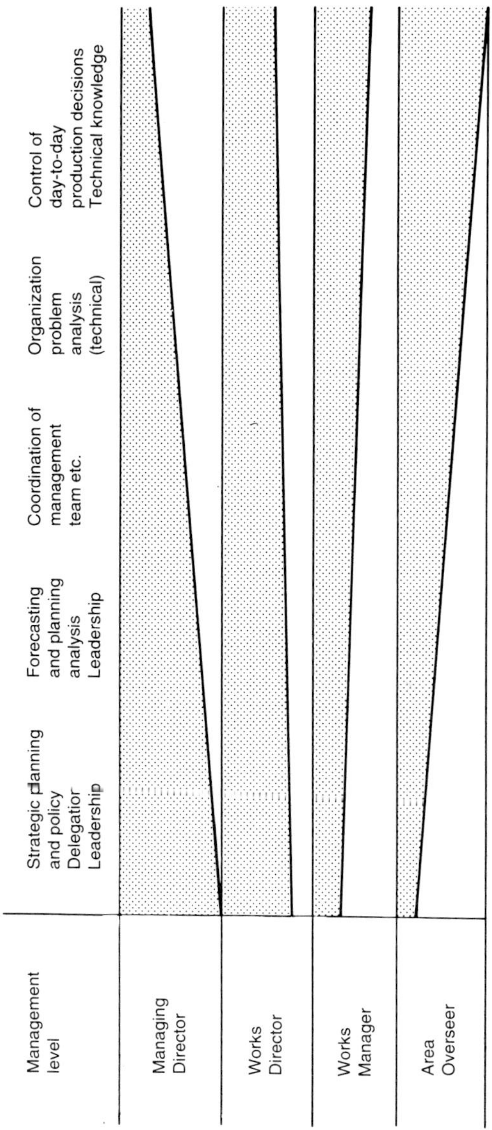

Figure 1.2 Skills required

- strategic-planning skills that assess future requirements and decisions relating to action to be taken,
- self confidence, i.e. a belief in one's ability to overcome difficult situations,
- motivational skills, entailing the exhibition of enthusiasm and the development of enthusiasm in others to further company aims and objectives,
- the ability to delegate, which is a key management skill (successful management is the ability to achieve through others, and it requires that power and authority be entrusted to another person).

Obviously, the demand and the implementation of these skills will vary with the level of management position. A production

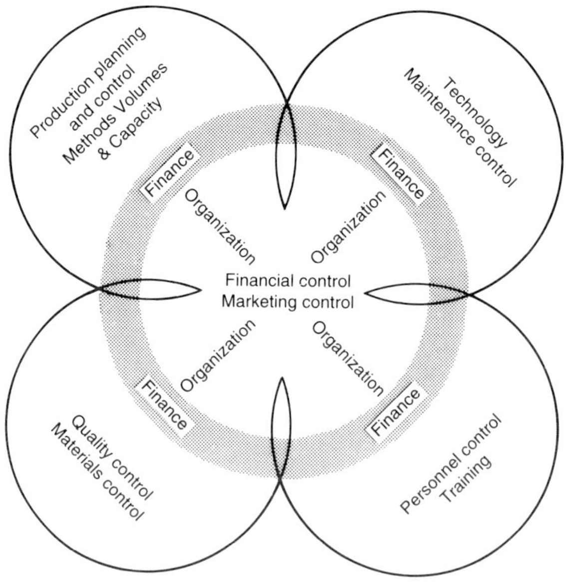

Figure 1.3 Interrelationship of management areas

overseer is far more likely to be involved in day-to-day technical decisions related to production, with perhaps a minor input to policy and strategy, while a managing director will be primarily involved in strategy, policy and financial decisions, responsibility for the day-to-day operating problems being delegated through the management structure. An analysis of skills in terms of the person's management position is shown in Figure 1.2.

The process of control and implementation cannot allow these skills to be viewed as isolated areas, for each skill has influence upon the others. It is only through the integrated assessment of the whole situation that any understanding can be reached or progress made towards managing the company. Figure 1.3 shows the interrelationships of the various management areas. All are dependent upon efficient integrated organization. Finance plays a part in all the areas.

TYPES OF BUSINESS

The various classes of business that are commonly found in the printing industry are as follows:

- *Sole trader*: This is the simplest type of business. The owner usually trades under his/her own name, although it is legally possible for the business to trade under a name other than that of the owner. Even if it does not trade under the owner's name, the owner is still liable for all the debts, since no limited liability exists. The individual is the owner and manager/operator.
- *Partnership*: The statutory rules that apply to partnerships are the same as those for a sole trader. In addition, each partner is responsible for all of the liabilities of the business. Should one partner amass debts that are unknown to the other partners, the others will be responsible for all of those debts should the first partner disappear. For partnerships an agreement should always be drawn up by a solici-

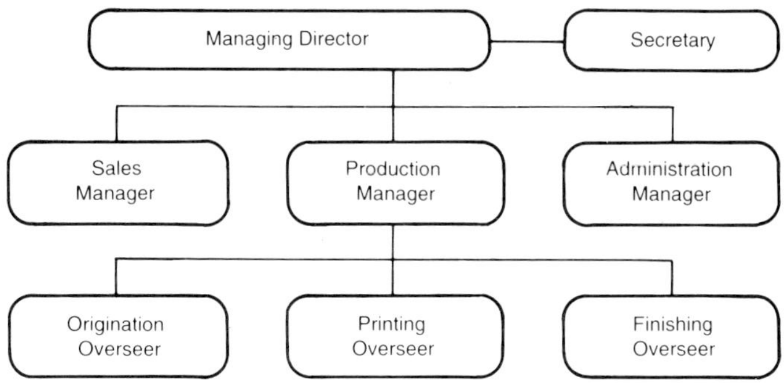

Figure 1.4 Management structure for small commercial printer

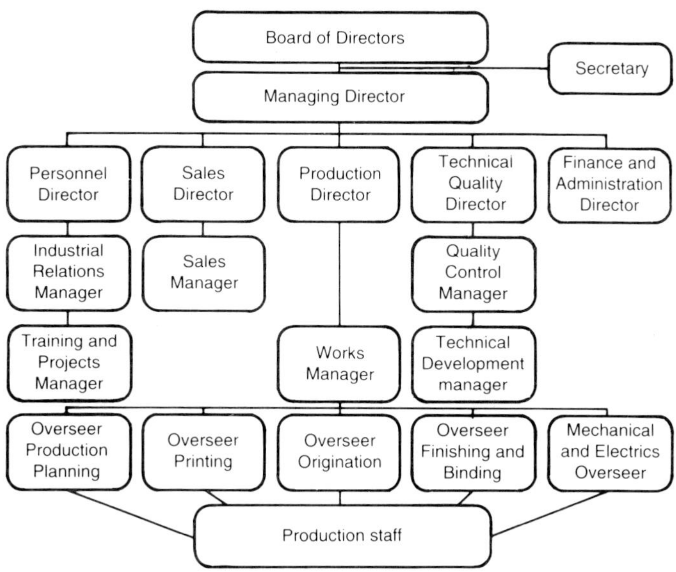

Figure 1.5 Management structure for large printing company

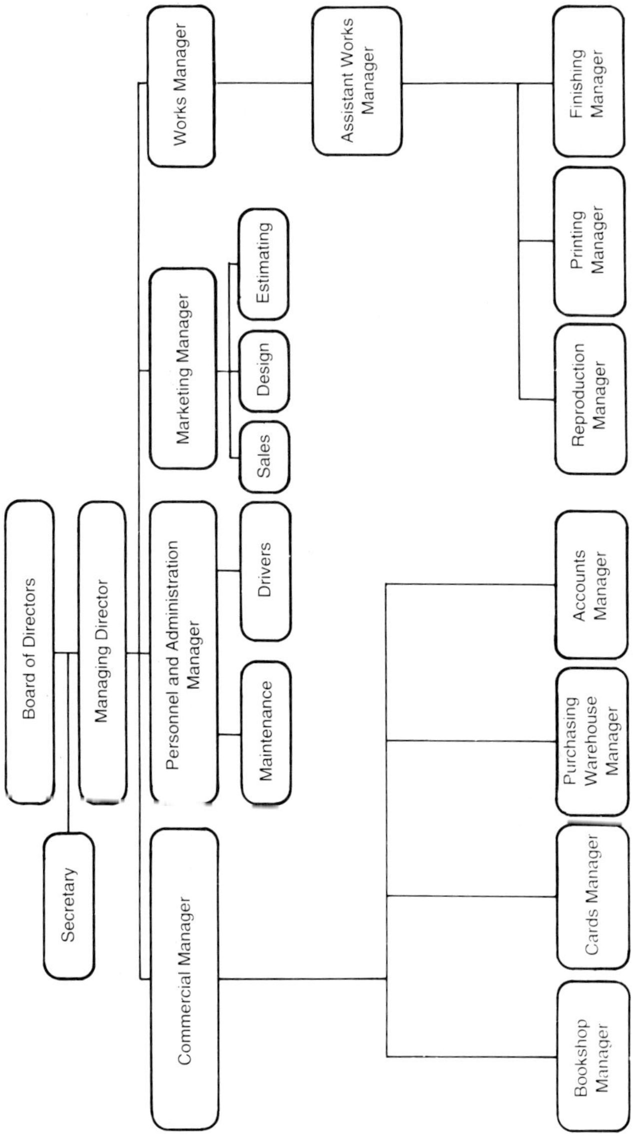

Figure 1.6 Management structure for printing company that prints books, card etc.

Figure 1.5 shows a structure for a large printing company. These companies usually employ 150–200 staff, and offer a wide range of production operations.

Figure 1.6 shows a structure chart for a large company that offers a wide range of operations, including design, and a book and card printing section. The wide range of services offered results in a more complex management structure, with each manager becoming area-specialized.

Figure 1.7 shows a structure used for in-plant printing operations.

The management organization charts show a line-management structure. This is the simplest type of organization, in which the company is divided into control divisions, such as finance or production. An important aspect of line management is that it clearly indicates the flow of authority from the most senior to the most junior employee. However, this type of organization can lead to inflexibility and autocracy. Also, interdepartmental cooperation can sometimes be difficult to achieve.

Another form of organization is often used that is known as staff or functional organization. In this, it is the function itself that determines the flow of authority, without there being precise operating divisions, as in line organization. A person will be placed in charge of a function, and will assume authority over that function wherever it may occur within the company. This type of organizational structure is very flexible, and is able to respond to rapid change. Demarcation is reduced, and function experts should be able to oversee the whole function.

There is a disadvantage, in that production workers may be left in doubt about whose authority is to be accepted. Line and staff organizations may be combined to benefit from the best features of both methods.

All management-structure and organization charts should show clearly the flow of authority and responsibility within the company, via a clear diagram of the various duties delegated, and the positions for which each manager or overseer is responsible.

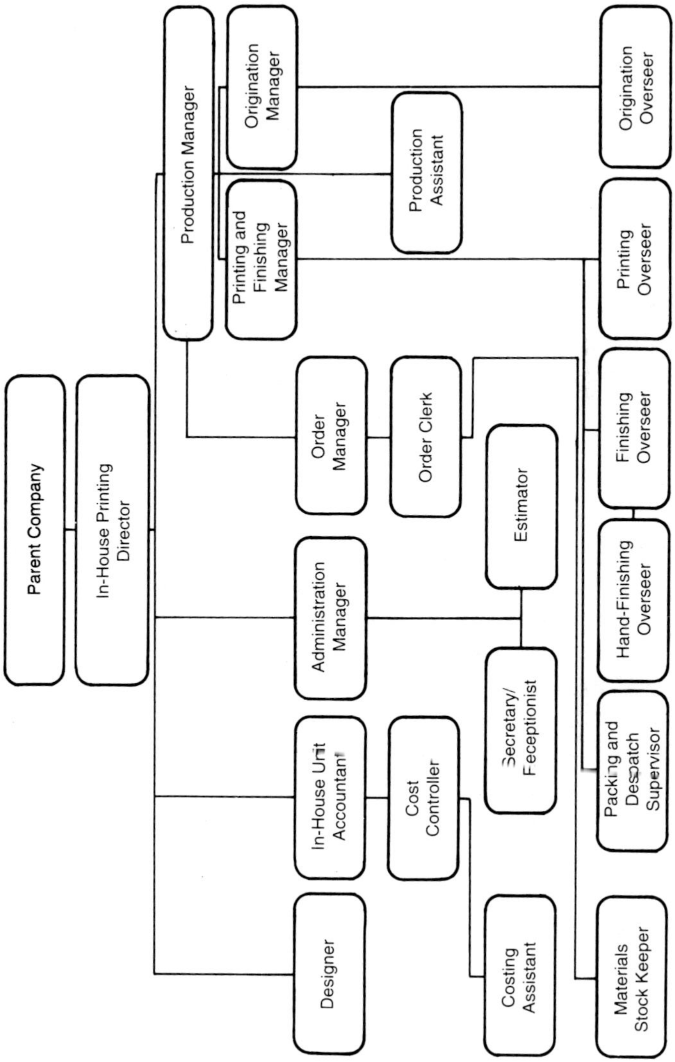

Figure 1.7 Management structure for in-house printing unit

The preparation of a management chart must involve careful consideration of each activity and administrative or production step required. Some distinct advantages will emerge, relationships will be clarified, and duplication of effort eliminated, and the channels of communication and responsibilities should be clear. Communications channels can be both vertical and horizontal. Vertical information usually flows downwards through a line-management structure, upward communication only occurring when necessary, when questions are answered etc. However, modern management has recognized the advantages of two-way communication and it is now widely accepted.

Horizontal communications take place between staff of about equal position but in different functions. Coordination must be achieved horizontally. Lateral relationships are those between equals, although they are not necessarily under the same supervision. Obviously, the ultimate objective is to form an effective management team consisting of skilled and efficient people working together in a coordinated effort. As Andrew Carnegie said, 'if I lost my factory and equipment, in fact all physical assets, even continuity of sales and financial states, but retained my organization, I would be fully reestablished in my own profitable operation in three to four years'.

MANAGEMENT-TEAM RESPONSIBILITIES

The responsibilities of the management team are as follows:

- *general administration*: consideration of future objectives and means of achieving them, organization development, salary management, industrial relations, training, working conditions, quality assurance,
- *marketing and sales management*: customer liaison, sales analysis, promotion, and advertising, market research,
- *production management*: production scheduling and control, quality control, maintenance, forecasting of technological

demands and trends, materials acquisition, movement and usage, warehousing, transport, production records etc.,
- *financial management*: managing capital, costing and estimating, pricing, credit control, accounting, insurance and property, company law.

BUSINESS PLAN

Without a business plan, the management will not know how to get where it wants to be.

Currently, few printing companies provide an overall service to their clients. The large financial investment required to purchase new technology, together with a complex range of staff-skills requirements, mean that there is an increasing trend towards small companies specializing in a particular area of print production. Trade origination (repro), specialist word processing, camera operation and scanning, printing by any of the major printing processes, by sheet and web, and finishing and binding, can all be found as individual specialist operations. Clearly, a company must decide on a number of issues related to its proposed production field and the expected progress within its specialization. This is usually indicated in the business plan. The plan covers the following:

- An introduction to the company's aims and objectives and current standing, including areas of specialization within the complex of printing-industry operations. A clear understanding of what the company can achieve without resorting to excessive subcontracting is advisable. However, specialization can often be a step towards greater efficiency.
- A guide to the industrial background in which the company is operating. An outline of the interaction of the business activities as a means of optimizing the performance of the business as a whole should be included, together with an appraisal of the general printing-industry situation, and those aspects of it that are specific to the company's opera-

tions. An evaluation of the competition in various possible trading climates is essential.

- A clear definition of the company's products. Very few printing companies offer a total production service, and therefore areas of specialization should be identified and promoted. An analysis of the implications of industrial trends should be attempted.
- The present and expected market share, including an analysis and forecast of existing and potential clients' needs in relation to existing production (e.g. one should not purchase technology simply because it is the latest available; customer needs, and therefore sales potential, should be researched, and then, if appropriate, the technology should be adjusted to meet demand).
- The structure and nature of operations, including an outline of operations, processes and the technology currently in use (e.g. DTP, origination, surface preparation, proofing, printing, finishing and binding; processes may include lithography, flexography, photogravure, screen processes, and foil blocking, and the technology includes the mechanical and other equipment in use).
- Employment policy and training arrangements, which covers the company policy on staff numbers, job evaluation, required qualifications and the provision of benefits such as health cover and insurance etc. An outline of the company attitude to ongoing staff training within the company and via equipment and machinery manufacturers and colleges etc. should be provided. The overall intention should be to develop a highly trained workforce for the benefit of company and employee.
- A management structure, i.e. an outline of the management flow of responsibility, usually drawn as a chart, showing the organizational structure via individual management positions.
- Financial data, in the form of financial statements about the current position and trading forecasts for 1–5 years. Detailed budgets can be included, depending on the re-

porting circumstances relating to the formation of the business plan. Various assumptions are made about the achievement of trading, profit and development objectives.

- The expected areas of development and improvement. Actions proposed as a result of an analysis of existing trading conditions and a forecast of likely future circumstances should be included.

The preparation of a business plan provides an excellent opportunity to assess any problems, strengths and weaknesses, together with any potential not yet realized. Business plans are formulated with the current and predicted future situations in mind, and contain a recommended course of action and a strategy to resolve problems and achieve objectives.

MANAGEMENT STYLES

Printing-company production can be substantially influenced by an individual manager's knowledge and perceptions about how to manage a company.

Aspects of management style include a manager's perception of his/herself, and his/her perception of the environment, and the manner in which these perceptions influence his/her behaviour towards people. It is essential to understand the kind of things, attitudes and circumstances which affect behaviour, and to be aware of the way in which these may affect responses to any situation. One may, for example, ask oneself, 'how do I react to change, or the need for change?' 'Am I uncomfortable when faced with change?' Only when one answers these questions will it be clear and understood how one will react to changes in printing technology, human and industrial relations, and production situations, and what one's perception of the company and its purpose is.

Individual beliefs about how the production workforce is persuaded, influenced, motivated and instructed to improve production has a critical impact on one's behaviour in achieving aims. It is true to a considerable extent that we obtain from people the productivity that our attitude leads them to assume is expected of them. What staff believe about working methods and processes will influence their attitude to change. In this context, proposed change, administrative or technological, must be practicable. The main management skills and methods are as follows:

- *Conventional (traditional) management*: conventional management solves its problems mainly by trial and error. Little data collection, evaluation and analysis occurs.
- *Systematic management*: systematic management is mainly based upon experience gained from watching and imitating other managers. This type of management has been popular in the printing industry in the past. There are serious disadvantages:

    - bad practices or unsubstantiated opinions can be handed on to successive managers,
    - initiative and innovation is discouraged.

- *Scientific management*: scientific management is based upon demonstrated facts rather than rule of thumb and individual opinión. It is based upon accurate measurement and control, and it therefore involves research and standardization. High investment costs and changing human-resources skills in conjunction with reduced profit margins have resulted in considerable attention to management style in relation to achievement in the printing industry.

Management styles vary in terms of the relationship of the manager to the worker with respect to authority and obedience, from the situation in which the manager exercises authority in the extreme, thus adopting an autocratic style, to that in which the manager is over-cautious about upsetting a worker, thus adopting a subordinate-led style of management. An autocratic style of management assumes that a worker must do as he/she is told, and no more and no less. This attitude is almost always counterproductive, and it can result in a worker using his/her knowledge and ability against the company, rather than for it. An opposing style allocates work to staff, leaving them alone, and letting them work as they see fit, handling technical problems in any way they wish. This system has led to the demise of printing companies, especially in the recent years of rapidly changing technology. Technological innovation and advances are ignored in this style of management, owing to the need for positive decisions, and the avoidance of problems in human relations.

Each management style demonstrates definite but differing perceptions of how to manage situations of production involving people. It is necessary to assess factually the variables between the two extremes of authoritarian and subordinate-led management, and decide upon the management style that is the most likely to achieve maximum production.

The appropriate management style for any manager can, in various circumstances, be determined by one condition or several conditions in conjunction.

Management styles and behaviour can be further classified as follows:

Psychologically close manager:

- Manager's behaviour is more concerned with good human relationships, often at the expense of efficiency; does not seek to formalize relationships,
- is inclined to select friends from colleagues in the company,
- prefers informal discussion with staff rather than formal meetings,
- will only delegate on minor matters,
- is inclined to dominate subordinates.

Psychologically distant manager:

- Tends to formalize relationships, both with superior and subordinates,
- accepts and rejects subordinates on the basis of performance,
- has a preference for formal consultation rather than seeking opinions informally,
- tends to be somewhat reserved and withdrawn in relationships within the firm,
- prefers ambitious subordinates,
- expects subordinates to make mistakes, and plans accordingly,
- does not develop deep friendships with colleagues,
- expects considerable freedom of action from superiors.

The most efficient management style for printing companies is as follows:

- The manager formulates ideas on the basis of proven facts.
- The manager presents ideas and facts to staff, and invites questions and comments.
- The manager amends ideas and plans in the light of any unforeseen but important points that have been overlooked or misunderstood.
- The manager communicates to staff the final decisions about the relevant management issue.

Adopting this style does not mean devolving decisions to others; decisions quite clearly remain with the management and become its ultimate responsibility. The manager adopts a philosophy in which the nature of people and the best manner in which they can be managed is taken into account, to the benefit of company and staff alike.

Obviously, not all decisions require long and concentrated thought about the methods of approach and action. However, even the most straightforward production operations are influenced by management style. A completely autocratic style will almost certainly mean that the worker only does what is necessary when it is necessary. A manager who clearly remains in control but approaches the worker in a spirit of cooperation will certainly achieve greater production.

In addition to the question of management style, and dependent on the management level, there is the issue of technical knowledge. To gain cooperation and respect from the production workforce, a manager must display competence in technical knowledge of the production operation being managed. This does not mean that the manager should be capable of physically operating all the technology and processes, but he/she must have a knowledge of the operation together with the key issues in the successful operation of the technology.

MANAGEMENT DECISIONS

The ability to make sound decisions is a crucial factor in effective management. The success or failure of a printing company can be considerably influenced by the nature and timing of decisions. The necessary elements of decision making are:

- a sense of the aims, objectives and purpose of the company,
- a clear understanding of the problem or situation,
- sufficient information to enable a sound decision to be made,
- an analysis and synthesis of the three points above.

All companies will require decisions to be made that relate to:

- the company structure, aims and objectives, and the work field etc.,
- finance: capital investment in technology, budgets, recovery of costs, profit margins etc.,
- human-resources provisions: skills requirements, salary levels, training etc.,

- marketing and sales: the existing market and the market potential, pricing, sales promotion etc..

These main company-management areas can be broken down into many others in which decisions are required on long- and medium-term issues together with those requiring immediate attention. Any sound management decision is based on a number of factors. The decision-making process consists of the following tasks:

- Examine the situation, and evaluating it for its importance and its implications for other areas.
- Clearly define the aims and objectives. All the situations entailing decisions must be resolved within these parameters.
- Collect as much relevant data as possible, together with any opinions of others if appropriate or necessary.
- Analyse the facts obtained from data collection to establish the true situation. Remember that data may vary because of the time at which it was collected, and the experience and attitude of the collector.
- Consider all possible options after an analysis of the data collected. Select only the options that will achieve the objectives having due regard to:

  - the strengths of the situation,
  - the weaknesses of the situation,
  - the best possible outcomes of the action,
  - the pessimistic view of the possible outcome.

- Make the decision. Disseminate it through the established lines of communication as appropriate.
- Implement the decision. Monitor and evaluate progress.

Implementation will be made considerably easier if a plan similar to that shown in Table 1.1 is adopted.

Table 1.1 Action plan

Situation	Action required	Action by	Expected outcome	Date to be completed
Inadequate customer liaison resulting in invoice errors	Increase administration knowledge of artwork criteria and repro through training	Office Manager	Reduction of invoice errors, and thus reduction in wasted administration time	
Incorrect repro and artwork reception	Artwork reception through repro manager or works manager for short term	Works Manager	Improved client relations	
Prolonged estimate turnround and delayed invoicing resulting in work or finance loss and poor cashflow (most operations carried out manually)	Integrate estimating, costing and invoicing via installation of a computer system, thus creating a centralized information system	Office Manager	Creation of up-to-date database Improved financial reporting and control provides first-level analysis of cost recovery and evaluates added value	
Client complaints about deviations from job specification	Enter all details required for production on job bags Operate, supervise and maintain a quality control system	Office Manager Order Clerk Works Manager	Reduction in spoiled work and customer complaints Improved production attitude and less downtime	
Late delivery of work causing customer dissatisfaction	Develop efficient admininstration/production and administration/sales liaison, thus creating a sales/production link	Sales Manager Works Manager	Improved delivery times resulting in increased customer satisfaction	

Management teams are under continuous pressure to improve the company's performance. The adoption of a plan that is similar to that shown in Table 1.1, but that incorporates aims and objectives and the expected timescale for achievement, will concentrate thoughts and actions, thus reducing demands on manager's time through:

- helping to prevent errors and avoid stumbling blocks that may prevent the company from achieving its aims,
- providing guidelines that prevent unwanted or extreme changes,
- encouraging initiative to achieve objectives.

An example of such a plan is shown in Table 1.2.

Table 1.2 Basic action chart

Aims/objectives	Action	Timescale, weeks	Action by
Purchase technology to allow move from single-colour to multi-colour printing, thus accommodating	Evaluate performance of existing technology and that of proposed technology	2	Works Manager
client trends towards increased colour production	Identify capital required, method of acquisition, effective production hours, rates etc.	3	Managing Director Financial Director Works Manager
	Identify location, building-fabric alterations if required, installation and maintenance costs	2	Works Director Electrical and Mechanical Engineer
	Identify any (in-company or external) training requirements	1	Works Director Production Manager
	Coordination meeting, review and redefine if necessary	By week 8	All concerned

Management by objectives consists of a series of stages as follows:

- Establish the purpose of the company in terms of the categories of work to be undertaken, both present and planned for the future. Establish methods for assessing performance.
- Review the company structure in relation to the achievement of the aims and objectives. Revise the organizational structure as necessary.
- Ensure that the manager sets the objectives for him/herself, and those for his/her subordinate staff.
- Develop consensus agreement on objectives.
- Analyse the results related to objectives at various intervals.
- Adjust or remove objectives in the light of the results.
- Collectively analyse the objective achievements against the targets.
- Amend the objectives as necessary.

COMMUNICATIONS

A good formal communications system is essential for all printing companies. Without it, all attempts to carry out the business of the company will fail. Effective communications between printer and client, client and advertising agency, and studio and printer is crucial. Also, communication between those carrying out the various production processes is essential if correct administrative procedures are to be carried out and production efficiency achieved.

Good communications are the responsibility of everyone within a company. However, the key links in communication are the line manager and the workers' representatives within the management structure. Even so, communication must be a two-way process, moving up through the management structure as well as down.

Communication must be through the correct source to the correct destination, or otherwise information will reach those for whom it was not intended. Important decisions may not be made, or action may not be taken, because information has not been received at the right time, at the right place, or by the right member of staff. The vertical and horizontal communications channels are defined by the company's management structure. The bypassing of any stage in the route can lead to misinformation and misunderstanding, and probably management and worker dissatisfaction at different levels within the company.

There are four major forms of communication, as described below.

Verbal communication

The use of the spoken word, face to face, is perhaps the most effective communications process possible. It allows the rapid exchange of information, and body expressions through which the receiver can judge meanings and attitudes. However, verbal communication is far from error-free, as the receiver of a communication may perceive the message as he/she wishes it to be, or, alternatively, completely misunderstand the communication. Issues of print production specifications simply cannot be left to verbal communication alone, as mistakes are almost certain to occur.

A large amount of verbal communication takes place in the printing industry. A danger arises when verbal discussion or details are not supported in writing. Information stored in the memory alters according to the time that has since elapsed and the current situation. Good communication means that instructions or messages are understood and accepted by all. This is difficult, if not impossible, if only one's memory of a client's telephone call or other verbal communication is relied upon.

Written communication

Staff are constantly engaged in producing written communications from formal printed reports to handwritten memos. The way in which a written message is prepared and presented is critical. Any instructions must be written in clear and unambiguous terms if misunderstandings are to be avoided.

Print production is a complex operation, and, if errors and disputes are to be eliminated, sufficient information must be provided in written form for each job undertaken. Matters that cannot be left to verbal communication alone in a technical work specification are as follows:

- *paper and boards*: quality, weight, size, caliper, grain direction, time allowances, special makings etc.,
- *composition*: copy preparation, method of setting, page make-up etc.,
- *reproduction and platemaking*: originals, artwork, transparencies, colour separations, planning, proofing etc.,
- *printing*: process (sheet or web), extent, colours, ancillary operations, self colours, lamination, varnish etc.,
- *finishing and binding*: cutting, folding, adhesive binding, numbering, delivery methods etc.

The list is not exhaustive and there are many more issues, but it illustrates some of the areas in which written instructions are necessary. Considerable annoyance and lost production time can occur when communication is poor.

Meetings

Meetings are only effective when they are held for a specific purpose. Care should be taken that meetings do not become 'talking shops'. Most of the important business should have been achieved prior to the meeting, and thus the reaching of decisions becomes the meeting's purpose.

- Company aims and objectives must be clearly defined.
- Areas of trading or specialization must be clearly defined.
- Management duties can be divided into 14 main categories, some of which are common to all types of business.
- No company can succeed without considering its staff and its management style.
- Management organization and structure charts arc essential to clarify lines of authority and responsibility.
- The evaluation of skills in relation to functions is essential.
- The identification of trading methods and structures is necessary.
- A business plan is desirable.
- Management style has considerable effect on overall company production, and it therefore requires identification.
- A decision-making process is required.
- Action plans will assist in achieving objectives.
- Good communications are essential to smooth and efficient operation.
- A management information system will allow management to be based on facts, and management to respond more quickly to required adjustment or change.
- Management self-development is essential to progress.

ACTIVITIES

- *Activity 1.1*: Outline the key organizational issues for a printing company.
- *Activity 1.2*: Discuss the advantages of the definition of management aims and objectives for a medium-sized printing company.
- *Activity 1.3*: Research methods by which a manager can progress through self-development.
- *Activity 1.4*: Identify an issue related to printing-company organization that is not discussed in this chapter.

REFERENCES

1 **Fayol, H** *General and Industrial Management* Pitman (1967)

2 **Drucker, P F** *Management — Tasks, Responsibilities and Practices* Butterworth–Heinemann (1992)

2
Print marketing and sales

Marketing and sales are regarded as one and the same thing by many printers. However, a difference exists. Marketing entails finding out what the customer requires, and selling entails persuading the customer to purchase from the company.

Marketing is a creative function, and it can be defined as everything that a company does to satisfy its customers, identify trading possibilities, and create satisfactory company profits. Marketing is an essential function of a profit-making company, and it will play a large part in its success.

The marketing function requires that the activities of production, sales, estimating, costing etc. be directed in a coordinated manner to finding out what a customer really wants. Marketing poses the questions of what customers require, and how it can be produced by the company. Marketing can also be perceived as everything that a printer does to create and maintain satisfactory levels of profit. Marketing and sales staff must work together to increase sales, and create overall trading policies for the company.

Selling can be defined as the personal contact of a salesperson with the customer and with the printing company. Personal selling is a vital function, and there will always be a need for people who can sell to others. In the printing industry, selling can be identified as providing the product that the customer wants at a profit that is satisfactory to the company.

This, at first sight, appears to be very similar to marketing, but a difference exists in that the salesperson will not knowingly be dealing with company planning and policy objectives, but with technical details relating to the customer's job in the context of price, delivery date etc. However, in some of the smaller printing companies, the marketing function becomes so closely related to sales that a distinction between the two functions is difficult to detect.

MARKETING

Marketing entails careful planning and therefore has strong *links* with, and makes contributions to, the company's business plan. There are three questions that must be asked:

- Where are we now?
- Where do we want to be?
- How do we get there?

From a marketing point of view, the answers to these questions will be found in the form of the aims and objectives in the company's business plan, which will include the following:

- A clear definition of the company's products. Very few printing companies offer a total production service, and therefore areas of specialization are identified for promotion. Often, an analysis of industrial trends is attempted.
- The company's present and expected market share. This includes an analysis and forecast of existing and potential clients' needs in relation to existing production.
- The existing structure and nature of operations, including an outline of the production operations, processes and technology currently in use, and any plans for new-technology installation in the near future.
- The management structure, showing the authority invested in areas or positions.
- Financial statements about the existing position and a trading forecast for the forthcoming year. Various objectives are listed in the form of targets relating to trading profit and

development objectives. Proposed actions resulting from an analysis of existing trading conditions, and an outline of the actions proposed to achieve the intended targets, are included.

Various issues must be considered to enable an understanding to be reached of the wide scope of the marketing function. These issues are often referred to as the marketing mix. The mix includes the following.

Promotion

Promotion is defined as market communication and liaison, and it involves the development of a reputation and the detection of any change in the company's image. Company product specialization, and the selling and advertising functions are also examined. It is important to remember that, no matter how efficiently a company can produce, it will not reach its potential if no one knows of its products.

A company must ask itself the following question. 'Why do our customers place orders with us?' 'Is it solely competitive price, or is it reliability, quality, or the standard of technology used, or good customer relations?' Often a successful company will answer yes to the majority, if not all, of the questions. Other questions that must be asked are as follows. 'What is affecting our market and our customers?' 'What area of the print market does the company's work come from?' 'Who are its main customers?' 'What are the customers' overall print requirements?'

Types of work

What type of work should the company offer? A starting point in determining this is to carry out an analysis of the turnover and profit by product type (see Table 2.1 for an example). Product types can, for example, be classified into the following categories:

Table 2.1 Example of record for analysis: turnover and profit by product type

Job type and description £		Apr £	May £	Jun £	Jul £	Aug £	Sept £	Oct £	Nov £	Dec £	Jan £	Feb £	Mar £	Total profit	Percentage
Multi-	Revenue	21 552.19	35 313.18	28 243.27	30 176.54	21 933.34	20 271.10	26 813.92	32 650.94					216 954.48	
colour	Costs	18 191.48	32 997.63	25 172.00	28 265.84	19 892.33	19 542.56	24 914.88	28 427.49					197 404.21	
booklets	Profit/loss	3 360.71	2 315.55	3 017.27	1 910.70	2 041.01	728.54	1 899.04	4 223.45					19 550.27	9.01
Single-	Revenue	23 034.00	32 576.00	34 220.50	57 649.00	59 893.00	68 702.00	94 664.00	73 411.00					444 149.50	
colour	Costs	20 170.91	28 221.21	29 111.18	55 437.71	58 033.75	56 857.11	82 199.08	64 387.69					394 418.64	
booklets	Profit/loss	2 863.09	4 354.79	5 109.32	2 211.29	1 859.25	11 844.89	12 464.92	9 023.31					49 730.86	11.20

- *point of sale*: showcards, display etc.,
- *posters*: small through to large formats,
- *stationery*: the whole field of commercial stationery,
- *magazines*: including periodicals,
- *books*: multipage single and multicolour,
- *packaging*: the whole field, through various processes.

Obviously, the areas of work will be influenced by the company specialization, but the principle remains the same. Identify the percentages of the various types of work, and therefore the operations within the company, and assess their contribution to turnover and profit.

An analysis of turnover by work type in relation to individual-customer profit margins usually shows that 80% of the company profits are generated by 20% of the projects, (Pareto's law). This is valuable information for the marketing function.

The analysis should also cover the quality of the products, and any production operations or process changes that are likely to occur in the near future.

Pricing

The pricing of products, especially within the printing industry, is difficult. However, a poor pricing strategy will have a marked effect on profits. Printing, as do other industries, has to cope with changing trading conditions, and fluctuations in labour and occupation costs. Thus, any marketing plan that considers pricing must incorporate an understanding of economic reality. For instance, major shifts in market location may lead to an adverse effect on profits. Market-location shifts may be because of intense competition, or simply a change of company location resulting from legislation, e.g. a move to an industrial area within a local-authority area.

Any examination and analysis of the types of work currently undertaken or considered for the future must include an evaluation of the technology available for production. It is pointless to attempt work that is unsuited to the company's existing technology, and therefore a careful evaluation of technology efficiency should be carried out. The selection of technology is sometimes based on feelings rather than on a true analysis of its capability and suitability for the type of work to be produced. Market research will provide a better indication of the technology required, and thus marketing should always commence prior to the installation of new technology, and not the other way round (in which a company installs new technology and then seeks work to keep it occupied).

Also, from the marketing point of view, we have to consider the fact that a customer purchasing printed work is not likely to be placing an order for the printing alone. The customer is really buying the work for what it will do for his/her business. For example, a customer purchasing a holiday brochure will evaluate the success of the work not only in terms of the quality of production, but also in terms of how many holidays are sold by the brochure.

The customer will not always buy at the lowest price. Print customers rarely buy on price alone, and therefore it is a mistake to assume that price is the sole motivator for placing an order. Quality, reputation, delivery times, and personal relations all play an important part in the overall marketing function.

Correct pricing can only be arrived at through the collection of data about production and other costs at different levels and operations. The prices charged by competitors should be taken into account when determining the company's pricing policy, but they must not become the key issue.

Various methods of pricing are in use.

Cost-based pricing

Cost-based pricing is based on the overall costs of production and administration, the overhead expenses, the marketing and/or selling costs, and a profit margin. A disadvantage is that it ignores competitors' prices. However, it has the advantage of setting minimum price levels, thus reducing the possibility of prices being set that result in financial loss.

Market-based pricing

Market-based pricing operates mainly on the basis of the customer perception of the 'going rate for the job'. This approach may result in the printing company developing an overoptimistic view of the value of its services, and it may therefore be more subject to the influences of changes in volumes of work. A general reduction in work volume often results in a reduction in prices, and this can be to levels at which profits are inadequate or nonexistent.

Competitive pricing

In competitive pricing, prices are set at a level below those of the major competitors. This can only safely be done when the actual costs of production are known. Should this system of pricing be adopted, extra care must be taken to establish the true costs of operation prior to any price reduction. An important consideration is that the setting of lower prices may restrict the company to poor-quality work, as the great majority of printing is not purchased on price alone.

Alternatively, prices can be set above those of the major competitors. Some customers will perceive high prices as an indication of superior quality and service.

Distribution

How will the products be delivered or distributed? This decision entails an evaluation of the channels through which the finished product will be distributed, e.g. directly to the advertising agent or customer, or through booksellers, publishers etc.

The great majority of printed work is in the form of direct sales, and it is delivered directly to the client or a nominated recipient. In some cases, a third party will be involved, such as a print buyer or consultant, and therefore the printer has little, if any, contact with the original purchaser.

Distribution may also be through agents, a system that is, often used in magazine, newspaper and periodical production. This method will involve a distribution chain consisting of the printer, publisher, wholesaler and retailer. Part delivery and work storage will also have implications for the company's marketing and sales policy.

Overall market direction

The management must have a favourable view of and approach to the marketing function before they can provide overall marketing direction. Marketing can be expensive, and it does not in itself guarantee company success, although it forms a vital part of company success. However, a marketing-oriented printer who understands that the prime concern of the company should be to satisfy its customers' requirements profitably will always be in a better position than a competitor who considers that any marketing function is unnecessary as sales and profits are adequate. All marketing should be profit-oriented and customer-focused.

Marketing requires financial investment, and it is sometimes considered to be too expensive, and it thus becomes subject to pressure to reduce costs. Marketing should be seen as an essential investment, and not as an avoidable cost. This does not imply that the marketing function should have unlimited funds. In fact, marketing budgets should be carefully considered, prepared and designed to meet business objectives.

Market research

Some printers assume that a continuing market exists for the work they produce. This is not always correct; changes in technology and improvements in production methods and processes often result in shifts in demand. This can be seen, for example, in the major increase in colour reproduction that has occurred over the past few years. This trend and many other trends will continue to develop and have implications for printers' shares of various markets. Prevailing economic conditions may see work that has been printed in one part of the country or the world being transferred to another. While this may not immediately affect a printer, there are still important issues of interest, such as those of what are other printers are up to, and whether they pose a threat to the company's operation.

Market research provides the following:

- An understanding of the print market as it exists at any one time.
- An understanding of how the industry has changed, and the reasons for change.
- A vision and forecast of possible future changes and how these are likely to affect existing markets. An example is the change in market opportunities as a result of the rapid changes in electronic-imaging technology. These changes have made many origination processes that were previously available only to experienced printers available to a much wider group of people. A result of this change can be seen in the move, in many cases, of origination operations such as film output to advertising agencies or publishers, away from the printer.

Market research commences with seeking answers to questions about the client base, work types, profit margins etc., and it continues through the research into and evaluation of production opportunities, and the company's standing and image.

Methods of research include the following:

- Questionnaires to existing and potential customers. The questions should not only relate to existing circumstances, but also to future potential. They may take the form of an enquiry into the installation of new technology, and the volume of work that may be expected should such installation take place.

- Personal contact and enquiries by the sales team. This can be the most effective way to evaluate future potential.
- Telephone discussion between sales managers and the customer. However, this method is useful only to establish later contact. A disadvantage is that one cannot show examples of the company's work over the telephone.
- An analysis of company records.
- Published information in the form of printed articles or records, together with project reports from Pira International, the British Printing Industry Federation etc.

An analysis of a company's market research will probably include the results of previous marketing action. These results can provide a useful base of information for future project planning. However, it is essential to ensure that any information is accurate, because otherwise the company's true position cannot be known. A SWOT chart (see Figure 2.1) can be used in the evaluation. The results can be used to develop a market-

Strengths	Weaknesses
Opportunities	Threats

Figure 2.1 SWOT chart

ing strategy, and provide an overall direction for the company's marketing and selling operations.

Market areas and specialization

The printing industry comprises many companies operating in various areas of print production, as follows:

- Printing companies that specialize in, for example, word processing, repro, printing, and finishing. Few printing companies offer a complete operation to the customer. The printing industry is made up of a large number of small companies, often employing under 20 staff, and concentrating on limited areas of production.
- Printing companies that specialize in customer categories or groups, e.g. publishers, finance houses, hotel chains, and customers dealing with art reproduction and limited editions.

Specialization can result in the development of unique services at competitive prices, because one printer does not cover all the specialist production operations. The identification of a niche for the company in the market is desirable. Unfortunately, specialization alone does not make a company profitable. There are many other issues, such as management competence, staff skills, and the technology used. Each customer type, or work category, provides potential market opportunities, but strong competition is faced also. Usually, a customer will give certain types of work to specific companies. This arises from the assumption or conviction that the company is equipped with the management and production ability to meet specific work requirements. An example is that of the specialist book printer. The publisher will select the printer because of an understanding of the printer's ability in this field of production. Often, the customer's decision is based on the range of technology and the methods of production used in terms of their suitability for the type of work.

Correct market research carried out by the printer will identify any trends or definite changes in production style or methods required by the publisher that may affect medium-to-long-term opportunities for the printer. Production specialization is often a step towards increased efficiency, and thus market potential. It allows:

- better use of the existing technology through a clear understanding of its potential, and reduced make-ready and setting-up times,
- increased standardization, and thus less waste,
- the easier achievement and maintenance of quality standards.

The case for operating in specialized areas is very strong. The large financial investment required to install the latest technology through a wide range of production operations is prohibitive to all but the large printers. The selection of a production area of the industry, and the development of expertise within that area, is more likely to result in success for the company than an attempt to provide an all-through operation.

Other issues to be considered are the following:

- The location of the printer in relation to his/her customers, i.e. the ease of liaison where personal contact is required.
- The ease of distribution and delivery, i.e. access to motorways, railway depots and industrial sites. Evaluate the possible delivery problems. It is no use organizing efficient production if the company is to experience delivery problems during the last stage of customer satisfaction.
- The costs of premises, taxes and maintenance. The company's location can have a great effect on the occupation costs.
- The costs and sources of materials, the reliability of acquisition, and the time taken to obtain materials.
- The availability of correctly trained and skilled staff is very important. Specialization must include the provision of

suitably skilled staff, which may not be as easy as it first appears, and it may require training and development. Marketing is a team effort, and it involves both management and production staff. All production staff should understand that the prime purpose of the company is to satisfy its customers at a profit, and therefore production and other jobs only exist as long as customers are willing to place work with the company. Production workers, through their skills, have a direct impact on the customers' perceptions of the company. It is often said that a printer is only as good as his/her last job for an individual customer. Any failing by the printer will probably be remembered more than all the successes.

The management attitude to customer care is important. Good results can only be achieved through correctly trained and motivated sales staff and other staff who are committed to aims and objectives that they value. Coordinated team effort is essential, and no one department in the printing company should be regarded as superior to another. Every member of staff has to complete his/her job to the best of his/her ability all of the time.

The company's marketing concepts and policy should be carefully communicated to all staff. Marketing and sales, in the same way as quality control, incorporates an attitude of mind. A correct attitude of mind towards sales and marketing is in everyone's interest for the future. Some printers exhibit their work and customer care policies throughout all the areas of the company as a constant reminder to all staff of the importance of this matter.

The majority of printers develop some sort of publicity to supplement the selling operation. Advertising is another form of marketing. Possibly the main form of advertising for many printers is their work, which in itself becomes an example of the work type and quality achieved. Thus the first criterion of informing the customer about the company's products is partially achieved. Advertising in relation to marketing can have a number of functions, as follows:

- to convince customers that any work placed with the company will be produced to their entire satisfaction using the correct production methods and technology,
- to inform potential customers about the range of services offered,
- to inform existing and potential customers about the technology in use and its specific purpose, as well as any planned future installation,
- to inform customers about the benefits to the customer of the company's service to the customer in relation to quality, price and delivery.

Printing companies will often produce advertising material that is designed to display a range of services or quality. The work may be in the form of a brochure that contains halftones with various screen rulings, or that uses various production techniques. One example would be a newspaper, printed on newsprint by web offset, with a complete range of colours that are similar to those of the Pantone matching system, but that have been printed using a four-colour process. These examples not only perform an advertising function, but are also a means of avoiding production misunderstanding or disappointment in the finished production.

Other printed advertising may focus on:

- customer service,
- competitive price,
- investment and growth,
- capacity and range,
- delivery to deadlines.

The success of advertising will be judged by the customer response and the interest manifested in terms of enquiries for additional or new work.

SALES

Definition of sales and responsibilities

A printing company's sales force is the first contact with its customers. It is often the sales force's skills and hard work that provides the ongoing workflow required to maintain production and achieve company aims and objectives. Printing sales staff need vision and a firm belief in the company and its products, as well as a conviction that the company's products are required not only by existing customers, but also by others. The salesperson will therefore seek out new work and opportunities on a continuing basis, and then sell to the companies in a manner that satisfies their requirements and that results in an acceptable profit for the printing company.

Sales promotion involves everything that contributes towards a customer placing an order. Marketing and sales activities will be influenced by the printing company's business plan. Therefore, the sales force must have a clear understanding of the company's aims and objectives, and be conscious of these aims and objectives when seeking out new business. Obviously, the aim should be to obtain work that the company can produce effectively with its existing technology. However a case for the installation of new technology can be made because of a constant demand for a class of work that the company does not currently undertake, or work can be subcontracted to other printers.

The successful achievement of sales entails the expenditure of time, thought, effort and financial investment, especially in times when increased competition has a great effect not only on sales, but also on profit margins.

Some people have a talent for selling. Others will find that experience assists their progress in this vital function. Successful sales is a combination of:

- personality,
- attitude,

- technical knowledge,
- a thorough understanding of the company and its products,
- experience,
- the ability to listen,
- a clear understanding of the customers' requirements, and the related price considerations.

Personality

It is difficult to define a specific sales personality, as no two people will sell in the same manner. The salesperson's personality should be outgoing (extrovert), and he/she should enjoy meeting people on both business and social occasions. Some of the most successful salespeople have a strong ego combined with a strong inward desire to succeed.

A salesperson should be able to overcome rejection philosophically, and to recognize and analyse the basis of customer resistance. An honest approach combined with integrity will always be appreciated by a customer, and, if it does not immediately result in obtaining work, it may well do so in the future. Perseverance over a long period will often pay off with an eventual sale. Sales do not just happen. Selling is an ongoing process that requires hard work all of the time.

Successful salespeople have far more than a pleasant manner and an ability to relate to people. They also have an indefinable ability to encourage people to listen to them, and, at the same time, to detect the potential for a sale.

Attitude

Concern and, to a certain extent, fear about the customer's reaction to the selling process is natural. These problems can, to a large extent, be resolved if adequate sales preparation is carried out as follows:

- Consider the prime objectives of the interview, meeting or customer contact, and keep to them.
- Be willing to listen to the customer's needs in preference to overemphasizing what the printing company has to offer.
- Maintain the customer's attention by talking in terms of the customer's interests.
- Prepare samples or presentations of the printing company's work in case they are required by the customer.
- Avoid being drawn into making promises that are unlikely to be fulfilled.
- Adopt the attitude that the company's competitors are good, but your company is better.
- Never argue with a customer, even if it appears that he/she is wrong. This situation may occur when discussing production methods or processes. Asking questions can often resolve potential problems.

Above all, a salesperson should understand the value of creating a good impression. He/she must understand human nature, as this element is often prominent in sales transactions. Autocracy and aggressive persuasion are unlikely to result in success.

Desirable personality traits for the salesperson are:

- stability,
- perseverance,
- the ability to mix well and get along with others,
- loyalty to the company,
- leadership.

Technical knowledge

Print production consists of a series of complex operations that use a wide variety of materials. A salesperson wishing to succeed should have a good technical knowledge of all the production operations carried out, and not only those carried out within the company. The salesperson's technical knowledge should be sufficient to allow intelligent and factual discussion with the customer about the work. However, care must be taken to avoid technical jargon unless the customer uses it first. Using jargon terms that are unfamiliar to the customer will have a distancing and demotivating effect upon the customer. Remember that he/she is unlikely to agree to something they are confused about. Things are even worse when the customer misunderstands a term used to describe an operation or process, as this may result in embarrassment, spoiled work and/or damaged customer relations.

The salesperson should have adequate technical knowledge about the key issues in production and the customer's work requirements. These include:

- the style and presentation of the artwork,
- the methods of design and origination,
- the suitability of the various production elements, e.g. transparencies, monochrome illustrations, and line drawings,
- the suitability of the chosen printing process, e.g. lithography, flexography, screen processing, and letterpress (an efficient salesperson should be able to detect quickly the process used for any piece of work),
- the suitability for their end purpose of the materials intended for use, in terms of substance, quality, surface etc.,
- the possibilities and effects of various colour combinations in terms of colours and materials.

Some printing companies select production staff to be trained as salespeople. In this situation, care should be taken, as technical knowledge alone is not a sound foundation for efficient selling. Many other issues are also involved. Being a good production worker with excellent technical knowledge is not sufficient for a person to become effective in a sales career. While technical knowledge is important, the ability to sell is even more important. Technical knowledge can be gained more easily through training than can selling skills.

Experience

Experience will create a confident and positive manner in the salesperson, and encourage the customer to believe in the printing company's ability. It will also enable the salesperson to present the company and its products in an interesting manner. An understanding of the company's history, potential, methods, and quality-control procedures will create a positive attitude in the salesperson.

Commercial knowledge

A salesperson has a commercial role, and therefore he/she needs to have commercial and business knowledge. A clear understanding of the UK *Sale of Goods Act* and the law of contract is advisable.

Conditions of order and standard trading conditions will have a direct impact on certain customers' work, and they will therefore require careful study.

Although they may not be under the direct authority of the salesperson, the issues of credit and its provision or restriction should be assimilated by the salesperson.

Human element and sales

In all sales transactions, the human element is a large factor. Salespeople need to think not only of the printed product and the facilities being offered by the printing company, but also of the views and requirements of the customer to whom they wish to sell. The emphasis must be on what benefits are to be gained by the user of the printing company's services.

A fast response to a customer enquiry will often impress the customer. Enquiries about price, delivery or technical considerations should be resolved as soon as possible, and, if delay appears to be likely, the customer should be informed immediately. Delays in providing the information requested by the customer may not only result in the loss of one particular job, but of others also, as the customer will find the response time too long, and will place work elsewhere.

An ability to listen attentively and comment intelligently is very important, as it also develops flexibility of thought and the ability to react to changing circumstances within the sales discussion.

Price problems

Unfortunately, problems with print prices are common, and these are not the worry of the salesperson alone. Methods of production, technology, materials, and profit margins all affect the prices asked of the customer. From a salesperson's point of view, the type of customer must also be evaluated. It may be that the customer cannot afford the price asked, or that it exceeds the customer's budget for the work.

Price problems include the following:

- A price discussion may result from the customer's business policy, because of which a price negotiation is attempted in an effort to reduce price without loss of quality or quantity.
- There may be a price comparison with other printers supplying the customer.
- There may be a misunderstanding about the amount of work involved in an individual job. This is far more common than one would at first think.
- The customer may simply want something cheaper.

Experience supported by training will enable a salesperson to detect quickly the reason for price problems, and take steps to resolve the situation.

Motivation and remuneration of salesperson

An efficient salesperson is creative and self-motivated. People can be self-motivated, but motivation will depend on many conditions. Managers will make assumptions about people at

Table 2.2 XY Theory

Theory X	Theory Y
Man dislikes work and will avoid it if he can	Work is necessary to man's psychological growth
Man must be forced or bribed to put out the right effort	Man wants to be interested in his work and, under the right conditions, he can enjoy it
Man would rather be directed than accept responsibility, which he avoids	Man will direct himself towards an accepted target
	Man will seek and accept responsibility under the right conditions
	The discipline that man imposes on himself is more effective and can be more severe than any imposed on him
Man is motivated mainly by money	Under the right conditions, man is motivated by the desire to realize his own potential
Man is motivated by anxiety about his security	
Most men have little creativity, except in terms of getting round good management rules	Creativity and ingenuity are widely distributed and grossly underused

work, one extreme of this being McGregor's XY theory[1] (see Table 2.2).

Initiative and creativity in the sales force is preferable to the mere carrying out of an operation, as sales efficiency, and therefore profit, is not dependent on skill alone, but also on the utilization of individual creativity, energy and perseverance. Care must be taken to differentiate between motivation and incentive. Motivation can be encouraged by external actions, but it must also spring from within the individual, and it may well remain in force if the external action is withdrawn. Incentives are the forces applied by such people as sales managers to encourage achievement. Obviously, the discerning sales manager will motivate the staff as far as possible, as it will result in a higher degree of utilization of all the company's staff through the acquisition of work.

An examination of Maslow's theory of motivation reveals that people are motivated by five basic needs, the primary and most basic need being that of physical survival, and self-fulfilment being the highest level of need (see Chapter 4: Human resources).

The reputation and status of a printing company is another motivating factor that may not be entirely connected with remuneration levels. The effective motivation of individuals in a sales force depends also on management understanding of the sales operation. Ross E Smith's model of human-resources management identifies factors that affect individuals and their behaviour at work, as follows:

- organizational communications,
- the decentralization of authority,
- the use of management targets,
- the use of job enrichment,
- the ensuring of adequate combinations,
- staff being allowed to participate when they have meaningful and useful contributions to make to the solution of problems.

Remuneration

Various systems of remuneration for salespeople are used in the printing industry. Three of these are:

- remuneration based on profitability,
- salary only,
- salary combined with incentives.

Remuneration based on profit

Under this system, a salesperson receives a percentage of the profit for each job. However, he/she also suffers any loss incurred through spoiled work, even when the spoilage was not directly his/her fault.

This system tends to produce closer customer liaison and control of work in progress.

Salary only

This system is sometimes used with new entrants to sales, especially with young people just commencing their career in selling. It has the advantage of the salesperson not having to worry about potentially large variations in salary until they have become established, when a transfer to a salary-plus-commission basis will provide additional motivation.

Salary and commission

This method of sales remuneration is probably the most widely used within the printing industry. With this method, a salary is usually paid every month, and a commission based on the volume or value of sales is added. This method is popular with salespeople, as it provides a regular income, but also provides the opportunity to increase income through successful sales.

Sales managers also favour this system, as it allows control, and motivates the sales force. The system may allow commission on all sales, or commence at an agreed turnover or profit level.

Usually sales people have the use of a car at the expense of the company and a petrol allowance, depending on the extent of their sales territories. Other expenses, such as hotels, subsistence and exhibitions, are usually paid against receipts. Careful attention should be given to the salesperson's level of expenses in relation to the value of their sales.

Sales-profitability checks

Regular analyses of profits in relation to the type of work introduced by each salesperson should be carried out. Obviously, the aim is to establish a client base that allows the achievement of adequate profits.

The analyses show:

- the work achieved by an individual salesperson by type or category,
- trends in price levels,
- the salesperson's contribution to the overall company turnover and profits,
- the clients' importance and priority ranking.

This information, when analysed with marketing data, will assist in the evaluation of the existing types of work and their suitability for the existing technology, and the production methods and processes in use. Further analysis will reveal the percentage of work introduced to the company that requires subcontracting to other printers or specialists. From this, the sales force can be guided towards the type of work required, or the company can consider the installation of new or additional technology.

Sales organization

The primary duty of sales management is to provide work that is suitable in terms of the class of printing that the company is equipped to undertake, the technology, and the skills of staff employed.

Salespeople are usually organized on an area basis. This has the advantages of the salesperson's proximity to the customers, thus allowing personal contact if necessary, lower travelling expenses, and simplicity.

Only in the largest of printing companies do salespeople become responsible for a product group. However, this does have the benefit of in-depth specialization in the product and its benefits.

Alternatively, salespeople may be allocated to various customers by customer type. This may be the result of customer specialization, the size of the account, or personal considerations.

A salesperson's working time cannot be assessed only in terms of the time spent on recorded calls alone. Time spent in careful preparation is as valuable as customer meetings. Above all, a salesperson must be self-motivated and be able to work without constant supervision.

SUMMARY

- Marketing and selling are often regarded as the same activity, but a difference exists.
- Marketing has a wide scope that covers everything that a company does to improve its workflow and profitability.
- Marketing and sales targets are important parts of a company's business plan.
- The identification of work type in relation to turnover and profit is essential.
- Pricing is a difficult operation that can only be successfully achieved through the analysis of data.
- Various types of pricing are used in the printing industry.
- Establishing an overall market direction is very important.
- Market research helps to establish the future sales potential.
- Market-area specialization is common in the printing industry.
- Marketing and selling are a team effort.
- The management attitude to sales is important to success.
- Printers use various advertising methods to promote the company, as well as their own work.
- It is essential that there is always a sales promotion plan, and not just when work is difficult to find.
- The personality of the salesperson is very important in successful selling.
- The attitude of the company and its sales force is also very important in terms of sales.
- Technical knowledge is important to the salesperson, but not as important as sales ability.

- Experience creates a confident and positive manner in the salesperson.
- Commercial knowledge is essential for the salesperson.
- Problems of price often occur, and therefore the salesperson must employ skills to resolve the situation without dispute.
- Motivation and remuneration methods are important considerations for the sales manager.
- Sales-profitability checks analysed by salesperson and customer are necessary.

ACTIVITIES

- *Activity 2.1*: Identify the differences and links between the marketing and sales operations.
- *Activity 2.2*: Evaluate the criteria involved in market research.
- *Activity 2.3*: Explain the term 'marketing mix'.
- *Activity 2.4*: Identify the various product and market sources of a typical company's sales and profits, and assess their relative contributions.
- *Activity 2.5*: Identify a sales issue that is not covered in this chapter.

REFERENCE

1 **McGregor, D** *The Human Side of Enterprise* McGraw–Hill, USA (1960)

3

Financial aspects of print management

The success of any printing company depends ultimately on the efficient control of its financial resources. Printing companies far more often find themselves in difficulties because of finance than for any other reason. Printing production requires a very large financial investment in printing technology, and often in suitable property. The effective control of finance is therefore a crucial issue for the printing manager, as any person who has control of financial resources, has considerable power within the company.

Finance and professional accounting are highly specialized subjects. Even so, a print manager should do everything possible to understand the key issues involved in successful financial control. The complexity of print production, rapid advances in production technology and methods, and increasing competition make it more essential than ever before that printing companies establish effective financial control over all their operations.

This chapter develops an understanding by the print manager of the key issues related to financial control, and their implications, not only for the print manager, but also the company as a whole.

FINANCIAL-CONTROL DUTIES

The financial control of a printing company should be active and constructive, and it should be exercised by a financial director or manager, depending on the size of the company. Financial-control duties consist of:

- budgetary control,
- the management of working capital,
- the preparation of annual accounts,
- the maintenance of the financial records required in the UK by HM Customs and Excise, the Inland Revenue etc.,
- accounting,
- estimating and costing,
- the location and management of sources of finance,
- capital project planning,
- management accounting, i.e. the analysis of accounting records and costs, the identification of variables etc.

The advantage of the records produced in the process of financial control does not lie only in the discipline of preparing the records, but also in the value of the information that becomes available for use in the control of the company. Although financial records are historical, they still provide valuable information about trends against budgets, and they identify situations in which action is required. It is necessary to ensure that financial forecasts are as accurate as possible, and that any variance between estimates and actual costs are quickly detected. Management accounts present information in a way that assists the company's management in the formation of policies. It includes the analysis of cashflow, and information about the utilization of resources, overheads and liquidity.

Control is achieved through the following:

- *Management*: Financial management and control is a vital company function, and authority should only be delegated to those who are qualified to exercise it.

- *Authority*: The preparation and approval of orders, cheques etc. should follow clear lines of procedure and authority. Often, two signatures are required for cheques and very large orders.
- *Segregation of financial functions*: This helps to prevent mistakes or the possibility of fraud. Crossreferencing also helps in the determination of the accuracy of the records.

The main purpose of financial control is to make provision for:

- the meeting of all business expenses,
- the purchase of all the materials used in production,
- the payment of wages and salaries,
- the effective planning of future financial activities,
- the provision of information to shareholders and the company board about profits and losses, cashflow, and returns on investments,
- the assessing of asset replacement values,
- the carrying out of research and development,
- the planning of share issues.

An overall system of cost analysis and apportionment must exist before the company expenses can be successfully managed.

ASSETS

The assets of a printing company fall into the following three categories:

- fixed,
- current,
- intangible.

Fixed assets

A fixed asset has a specific value that will not be consumed during the course of company administration or production, for example:

- physical property, in the form of land and buildings,
- technology, tools and other equipment.

Any charges arising from these assets are difficult to reduce in the short term.

Asset replacement is essential when technology becomes obsolete, worn out, or simply unable to produce at the quality and volume required to maintain the company's position in the market and its profitability.

Current assets

Current assets include:

- material stocks,
- work in progress,
- working capital,
- cash in hand and at the bank,
- debts.

Intangible assets

An example of an intangible asset is goodwill. These assets may have a considerable influence on the stability and progress of the company.

FINANCIAL RECORDS REQUIRED BY LAW

In the UK, annual company financial accounts are required by the UK Government under the terms of the *Companies Acts.*

These acts are extensive, and they were amended at least five times up until 1989. Financial accounts are required in the form of:

- a balance sheet,
- a profit and loss account.

The balance sheet has to be prepared in a standard format, and it has to include the following information:

- the name of the company,
- the value of the sales for the year,
- interest payable (sources of interest have to be shown),
- the profit before tax,
- the taxation details,
- the profits after the deduction of tax,
- any extraordinary items,
- the retained profit,
- dividends to shareholders,
- administration expenses.

Copies of the balance sheet and the profit and loss account must be sent to all the directors, shareholders, and debenture holders of the company.

A profit and loss account must also be prepared by the accounts department in conjunction with the company auditors, in the same way as the balance sheet. The profit and loss account is a formal financial statement. The objective is to record all the expenses and losses incurred by the company against its gains, thus showing the net profit or loss for the financial year.

An example profit and loss account is shown in Figure 3.1. All the company accounts must be audited by a suitably qualified accountant.

The balance sheet is a classified summary of the company's financial balances, and it shows the situation on the given date (not just at the end of the financial year).

TRADING PROFIT AND LOSS ACCOUNT		
A B Printers	Period _______	
	£000s	£000s
Stock of materials at _______		300
Purchases		1096
Outside work		826
Transfer in		4
		2226
Less stock of materials at _______		274
Cost of materials used		1952
Direct labour		728
Prime cost		2680
Production overhead expenses		
Premises	182	
Renewals	274	
Indirect labour	128	
Other factory expenses	256	
		840
Add work in progress b/f at _______		148
		988
Less work in progress c/f at _______		218
		770
Total production cost		3450
Sales		4020
Gross trading profit b/d		570
Gross profit as a percentage of sales		14.2%
Administration expenses		
Salaries – print	182	
Corporate support	58	
Personnel	24	
Management	14	
Premises	26	
Renewals	6	
Insurance	2	
Other expenses	20	
Selling Expenses		
Distribution	38	
Marketing	14	52
Finance charges		
Debt charges	0	
Interest on stock	26	26
Total nonproduction costs		410
Net trading profit		160
Net profit as a percentage of sales		4.0%

Figure 3.1 Trading profit and loss accounts

A balance sheet cannot, in itself, be regarded as an account, but it is a statement of all the company's assets and liabilities on a particular date. It is a formal document with an accepted standard format that is prepared historically after the end of an accounting period.

An example balance sheet is shown in Figure 3.2.

BALANCE SHEET

Date________

A B Printers

Capital and liabilities	£	£	Assets	£	£
Capital	32 500		Land and buildings		25 000
Profit and loss account	4 500		Plant and machinery		12 500
			Vehicles		5 000
		37 000	Stocks		1 750
Loans		12 500			
Creditors			Debtors		
Trade creditors	2 500		Trade debtors	5 000	
Expenses incurred	500		Prepayments	750	
		3 000			5 750
			Cash		
			At bank	2 250	
			In hand	250	
					2 500
		52 500			52 500

Figure 3.2 Balance sheet

CAPITAL

The capital employed in a business is defined as all the capital that is available to the company, and it is treated as an asset. The capital consists of:

- funds provided by an owner,
- funds provided by shareholders or partners.

Working capital

An effective and sufficiently large source of working capital is vital for any printing company. The working capital may be defined as an asset set against liabilities. However, this may be over simplistic, as not all assets can be realized at short notice. Therefore, working capital should be defined as finance that can be drawn upon and utilized within a short period. Working capital can include the following:

- Stocks of materials, such as paper, board, film and plates.
- *Plus* finance owed by debtors, e.g. sums of money owed by debtors. Caution is required, as some debts may be difficult to recover. Customers who do not, or will not, pay on time will have a serious impact on the working capital. Materials suppliers will not willingly accept delayed payment. The printing company can be not only short of finance, but also without supplies.
- *Minus* all credit amounts. Credit is defined as all the finance that it is expected will be paid within the normal period of credit operated by the company. This will often be influenced by the state of the company's bank balance.
- *Minus* any overdraft at the bank. Overdrafts cost money, and, for a large overdraft, the interest may be considerable. This will obviously reduce the amount of capital that is available for normal operations or any intended capital project, and it may adversely affect any discount arrangements that the company offers.

Working capital has a direct connection with cashflow. Cash leaves a company in the form of payments to suppliers, payments of wages, and other expenses. Obviously, the company must have sufficient funds to cover these expenses. Otherwise, it may find itself in difficulties, or even in liquidation, even though it possesses considerable assets.

PRICING

The printing industry is a major service industry. Work is usually produced specifically for an individual customer, and it must therefore be priced accordingly. No two printers will estimate or quote identical prices for the same work. Often, the price differences are considerable. The variations in price can be due to:

- production methods or technology,
- company efficiency in relation to human resources and general management,
- the company's location,
- company production specialization.

Correct pricing is probably one of the most difficult, and yet vital, skills required in a company's financial management, and the financial manager will probably have an input. Any decisions about pricing will affect the company's profit and loss account. Print prices are influenced by the following factors:

- the price that the company needs to charge for its work to create sufficient profit for the company,
- the price that the customer is willing to pay for the work (this may be influenced by the price that the customer has been used to paying for similar work).

A pricing policy can be defined as the framework and rules used to make decisions about prices and their relationship to production. It can depend on:

- the business aims and objectives published in the company's business plan, e.g. profit targets in relation to turnover,
- the actions taken by the sales force to introduce new work and customers,
- general trading conditions in the printing industry, e.g. recession, and international competition,

Petty cash

A petty-cash book should always be maintained that records all the payments into or out of an initial float of cash. The minor sums of money used in day-to-day company payments are drawn from the petty-cash float.

Wages record

Any printing company that employs people must keep a record of the wages and salaries paid to its employers. In the UK, the record includes, for each employee:

- the national insurance contribution deducted from the employee,
- his/her gross earnings,
- the tax deductions,
- his/her net pay,
- the pension-contribution deductions,
- the national insurance contribution paid by the employer,
- the position held by the employee,
- his/her date of starting work.

It is vital that, in the UK, the requirements in law of the Inland Revenue and the Department of Social Security are completely understood. Both of these UK government departments produce literature that will assist the print manager to avoid problems.

From time to time, all the books and records are balanced. This activity is known as calculating a ‘trial balance’. From a successful trial balance, the following statements can be prepared:

- the trading account,
- the profit and loss account,
- the balance sheet.

These three accounts are regarded as final accounts, and many printers prefer to leave their preparation to a professional accountant. This is a very sensible idea, as the rules are complex, and are full of pitfalls for a company that lacks financial expertise.

VALUE-ADDED TAX

When a company is registered for VAT, it is obliged to record and collect or pay VAT on all purchases and sales transactions. In 1993/94 in the UK, all companies with a turnover of over £36 000/annum must register for VAT. This threshold alters from time to time according to government policies and directives. Keeping VAT records is mandatory, and it is advisable to keep all the records up to date, as late payment can result in a company experiencing cashflow problems arising from late-payment demands. VAT regulations are complex, and in the UK it is wise to seek advice from HM Customs and Excise.

NATIONAL INSURANCE

In the UK an employer has an obligation to collect National Insurance contributions. The sums of money involved can be considerable, especially in medium-to-large companies. It is important to remember that an employer has to pay a contribution for every employee, as does the employee. Unfortunately, it is easy to overlook this matter when carrying out financial planning, and this can have damaging effects on the company. Contribution rates vary from time to time, and it is wise to seek advice from the company auditor, or the DSS and the Inland Revenue.

SOURCES OF COSTS

The costs of a printing business fall into two categories:

- *Fixed costs*: These costs do not vary with the level of production, unless extra production technology is required to generate increased production. These costs relate to buildings, land, technology, and management salaries.
- *Variable costs*: These costs are directly related to the printed product. Examples are variable overheads that can be allocated to the product, and direct productions labour costs.

No company can afford to pay out more in wages, salaries, materials and other expenses than it receives in income from work produced or other sources. Company income must exceed expenditure. If it does not, the company is living off its capital, and, unless urgent action is taken to control the situation, the company may fail or be liquidated.

Although accountants are employed to oversee and advise on financial matters, financial data is nearly always obtained in retrospect. Therefore, an efficient system of cost recovery must be installed within the company. The system must not only recover all the expenses incurred in running the company, but must do so at a profit. The system must also allow all the financial functions to be managed day to day.

A costing system is thus used that records the costs of operation in each department. These costs include heating, lighting, power, occupation costs, costs of production, depreciation and insurance. The costs are then distributed between the various production operations, and divided by the number of productive hours that are normally worked within the company. The result is known as an 'hourly rate'.

In the absence of an efficient costing system, a company could make a profit and yet, at the same time, sustain a loss. This is possible if the profit margin is lowered from what it could have been by inefficient production methods or other forms of waste.

Costs of printed job

For the printer, competitive prices are a primary consideration. However, perceptions of value are influenced by supply and demand. Profit is what remains after all the costs incurred in producing a job have been deducted from the revenue. Printing companies price work to allow themselves to make a profit, and profit margins vary according to the nature of the work, the competition, and the efficiency of the company. Many of the areas and operations involved in print production are open to cost reduction. Areas in which reductions in costs can be achieved are:

- the prevention of waste through the efficient control of materials and working methods,
- the careful selection of the technology that is most suitable for the type of production undertaken by the company,
- the efficient use of human resources by the careful selection of staff on the basis of skills and flexibility.

The objective of cost reduction is to increase profits without increasing the price to the customer. Areas in which cost reductions are difficult are land and buildings. Reductions can usually only be achieved by moving the company to an industrial park or an authority-subsidized area.

Cost centres

All the operations carried out by a printing company are originators of costs, and any operation that requires financial investment and does not generate a return creates a financial loss. Cost centres include the following operations:

- design,
- origination,
- printing,
- finishing and binding,

- distribution, including transport,
- marketing and sales,
- costing and estimating,
- financial control and records,
- general administration.

Labour costs

Labour costs can be divided into direct costs and indirect costs.

Direct labour costs

These costs comprise all the labour charges that can be directly allocated to the product or customer. Methods of recording this information are as follows:

- *Daily work docket*: This docket is used in the manual recording of the actual time spent on each job by each production operator, and the time spent on the subdivisions of an operation, e.g. makeready and runon.
- *Computer keypad*: In this method, the production operators record the start and finish times of operations using a computer keypad. This method requires no simultaneous or retrospective handwriting. However, both systems must be used accurately if any useful information is to be obtained.

Indirect labour costs

Indirect costs arise from operations that are not directly attributable to a customer, such as downtime on machines, cleaning maintenance, and general assistance. Some of these costs relate to the department through which the work passes. The process of apportioning these costs can be complex and can vary. However, the main objective is to distribute these costs so that they can be recovered without undue pressure on any particular department or section of the company.

Materials costs

Materials used in production fall into two categories:

- direct materials,
- indirect materials.

Direct materials

Direct materials usually form part of the finished product. These include paper and board. All materials that can be directly allocated to a customer are direct materials.

Indirect materials

Indirect materials are materials that are used during the production process but are not directly allocated to a customer. Examples are processing chemicals, oils and greases, and press blankets.

Overheads

All printing companies incur overhead costs. These costs cannot be allocated to any specific job or customer. Therefore, it is necessary to distribute these overhead costs between the various production operations or cost centres. These costs include depreciation, repairs to technology and buildings, leave and sickness, holiday pay, and environmental costs.

Wages and salaries

The wage and salary costs for any production operation can be obtained from job records. The records include the hours worked by each individual in each department, and/or the hours worked by each worker on each job. This information can be recorded using a daily docket, as outlined above. Analysis by the accounts department will calculate the total costs of wages for production for any given period and area, and any overtime costs.

Department	Name	December 19 Basic	December 19 Overtime	December 19 Total	January 19 Basic	January 19 Overtime	January 19 Total	February 19 Basic	February 19 Overtime	February 19 Total	Totals Basic	Totals Overtime	Grand Totals
Camera	________	769.65	154.00	923.65	769.65	92.75	862.40	769.65	106.75	876.40	2 308.95	353.50	2 662.45
	________	917.50	168.00	1 085.50	917.50	189.00	1 106.50	917.50	152.25	1 069.75	2 752.50	509.25	3 261.75
	________	769.65	134.75	904.40	769.65	110.25	879.90	769.65	154.00	923.65	2 308.95	399.00	2 707.95
	________	696.35	265.24	961.59	696.35	191.32	887.67	696.35	143.07	839.42	2 089.05	599.63	2 688.68
	________	733.00	47.39	780.39	733.00	27.08	760.08	665.34	38.93	704.27	2 131.34	113.40	2 244.74
	________	439.80	0.00	439.80	439.80	1.02	440.82	439.80	9.14	448.94	1 319.40	10.16	1 329.56
	________	4 325.95	769.38	5 095.33	4 325.95	611.42	4 937.37	4 258.29	604.14	4 862.43	12 910.19	1 984.94	14 895.13
Printers	________	1 091.16	394.00	1 485.16	1 091.16	310.00	1 401.16	1 091.16	738.00	1 829.16	3 273.48	1 442.00	4 715.48
	________	974.25	152.00	1 126.25	974.25	272.00	1 246.25	974.25	464.00	1 438.25	2 922.75	888.00	3 810.75
	________	932.49	0.00	932.49	932.49	208.25	1 140.74	932.49	441.00	1 373.49	2 797 47	649.25	3 446.72
	________	439.80	21.32	461.12	439.80	43.65	483.45	419.50	54.81	474.31	1 299.10	119.78	1 418.88
	________	3 437.70	567.32	4 005.02	3 437.70	833.90	4 271.60	3 417.40	1 697.81	5 115.21	10 292.80	3 099.03	13 391.83
Finishers	________	956.93	242.00	1 198.93	956.93	340.00	1 296.93	956.93	158.00	1 114.93	2 870.79	740.00	3 610.79
	________	831.61	17.50	849.11	831.61	0.00	831.61	831.61	129.50	961.11	2 494.83	147.00	2 641.83
	________	549.75	11.41	561.16	549.75	48.17	597.92	549.75	53.24	602.99	1 649.25	112.82	1 762.07
	________	2 338.29	270.91	2 609.20	2 338.29	388.17	2 726.46	2 338.29	340.74	2 679.03	7 014.87	999.82	8 014.69
Admin	________	952.90	91.00	1 043.90	952.90	129.50	1 082.40	952.90	84.00	1 036.90	2 858.70	304.50	3 163.20
	________	623.05	106.38	729.43	623.05	115.00	738.05	623.05	120.75	743.80	1 869.15	342.13	2 211.28
	________	952.90	465.50	1 418.40	952.90	372.75	1 325.65	952.90	693.00	1 645.90	2 858.70	1 531.25	4 389.95
	________	144.00	0.00	144.00	167.40	0.00	167.40	207.90	0.00	207.90	519.30	0.00	519.30
	________	952.90	154.00	1 106.90	952.90	271.25	1 224.15	952.90	185.50 est	1 138.40	2 858.70	610.75	3 469.45
	________	769.65	108.50	878.15	769.65	87.50	857.15	769.65	96.25	865.90	2 308.95	292.25	2 601.20
	________	571.56	161.43	732.99	571.56	195.18	766.74	571.56	175.00 est	746.56	1 714.68	531.61	2 246.29
	________	623.05	0.00	623.05	623.05	14.37	637.42	623.05	37.38	660.43	1 869.15	51.75	1 920.90
	________	806.30	75.25	881.55	806.30	89.25	895.55	806.30	94.50	900.80	2 418.90	259.00	2 677.90
		6 396.31	1 162.06	7 558.37	6 419.71	1 274.80	7 694.51	6 460.21	1 486.38	7 946.59	19 276.23	3 923.24	23 199.47
Totals		16 498.25	2 769.67	19 267.92	16 521.65	3 108.29	19 629.94	16 474.19	4 129.07	20 603.26	49 494.09	10 007.03	59 501.12

Figure 3.4 Wages analysis

In the great majority of printing companies, the production efficiency of the workers does not affect the basic rate of pay. Therefore, it is vital that an accurate record of wages is kept, and evaluated in terms of the hours paid for, and the hours sold to a customer. The efficiency ratio of the company can then be calculated. An example wages record is shown in Figure 3.4.

Profit centres

Profit centres relate to the company's income and the profit achieved. They are often defined to be the same as the company's cost centres. However, the emphasis is transferred from costs to the income recovered from each operational area and the profit relating to this income. There are three types of profit:

- *Gross profit*: This is the financial surplus that remains after the deduction of the production costs and other costs of work sold from the sales revenue. The gross profit has not had the selling, distribution, financial and administration costs deducted from it.
- *Net profit*: This is the financial surplus that remains after the deduction of financial charges, e.g. overdrafts, interest and income from investments, from the operating profit.
- *Operating profit*: This is the financial surplus that remains after the deduction of the administration, finance, selling and distribution costs from the gross profit.

Costs and prices

The costs and price of a printed job are related as follows:

direct material costs + direct labour costs = prime cost

prime cost + production added costs = factory cost

factory cost + factory added costs = production cost

production cost + selling costs = cost price

cost price + profit = selling price

DEPRECIATION

Finance invested in capital items, e.g. technology, tools and buildings, is not easily accessible. All these items, especially technology, depreciate in value. (Buildings do depreciate, but land does not; in fact, it may appreciate considerably with the passage of time. Of course, for this to be of financial advantage to the company, the land has to belong to the company.) Therefore, when finance is invested in technology, although this is sound from the production point of view, it can have a considerable impact on the company's financial stability. Investment in electronic technology is a particular risk at the present time. This is because of the rapid advances in this area. Technology purchased to update the company's facilities may well be obsolete within two or three years.

Annual depreciation is calculated taking into account:

- the item's original costs,
- any resale or scrap value that the item has,
- the item's estimated useful productive life,
- the item's replacement cost.

Print managers should ensure that financial provision for the replacement of technology should be made through depreciation allowances. Failing this, finance for this purpose will have to come from income or other expensive financial sources, such as bank loans. Even with careful calculation, the depreciation rate is often insufficient to match inflation. Setting up a special reserve account may be one way of resolving this problem.

Some technology is certain to go out of productive use or end its life before the end of the depreciation period. Other items may continue in use long after their writing-off period. Therefore, it is wise to continue charging depreciation against all technology whether it has reached the end of its depreciation period or not.

The depreciation process allocates an appropriate percentage of the initial cost of purchase to each year of the expected life of the purchased item. For printing technology, this is 2–10 years; five years is common. A straight-line method of depreciation is most commonly used in the printing industry. This is shown in the following example:

initial cost of technology = £40 000
resale or scrap value at 5 years = £10 000
depreciation period = 5 years

$$\frac{30\,000}{5} = £6000/\text{annum}$$

Another method of depreciation is based on the principle that technology will depreciate faster in the earlier years, and that therefore depreciation should be calculated on a sliding scale to recognize this.

A record should be kept of all the technology. An example is shown in Table 3.1.

SOURCES OF FINANCE

The question of raising finance to cover the costs involved in replacing technology or other capital projects concerns all companies at some time. Sources of finance are usually classified into five main categories:

- short-term finance obtained through extended credit or bank overdrafts,

Table 3.1 Technology record

Description	Number of items	Reference or serial number	Year of purchase	Estimated life, years	Current replacement value, £	Comment
Roland PARVA-1 colour	1	12345	1985	10	140 000	
Heidelberg MOVPH	1	5434	1990	10	310 000	
Heidelberg MOZP	1	678910	1987	10	154 000	
Ryobi-1 colour	1	1601	1982	7	75 000	
Apple/Mac	1	1450455	1991	3	25 000	
Printing Down Frame	1	5421	1982	10	15 000	
Film Processor	1	16708	1983	5	9 000	
MBO Folder	1	45456	1987	10	80 000	
Guillotine	1	P19760	1972	10	62 000	Replace ASAP
Linotype Trim/Stitcher	1	L12354	1979	10	162 000	Replace within two years

- nonmoney-market sources, for example the retention of profits, and the reduction of materials held in stock and waste in production,
- long-term finance, obtained through loans and mortgages,
- medium-term finance, obtained through loans from finance houses, venture capital, hire purchase and bank loans,
- prompt debt collection, and the freeing of finance tied up in little-used technology.

Short-term finance can be a very expensive way of raising finance through bank loans etc. If a company is considering the purchase of technology, it is wise to consult and negotiate with the supplier first, as more favourable prices and payment terms can be obtained. Unexpected trading conditions or the need for financial investment can create cash problems, and thus careful preparation of the company's business plan is essential to reduce this risk as far as possible. The cost of raising finance within a company is calculated in terms of the return that the company must earn to cover all the liabilities to those advancing the finance. All printing production operations and projects carry some level of financial risk. Careful analysis of the feasibility of proposed financial projects will allow more than an intuitive guess to be made about the true costs involved.

COST/BENEFIT ANALYSES

When considering the purchase of an item of technology, a cost/benefit appraisal is advisable. This takes into account the capital investment and the ongoing costs, and the expected benefits. The benefits must be expressed in financial terms, for example cost savings. These can then be related to the capital costs to assess the return on investment, as follows:

$$\text{return on investment} = \frac{\text{cost savings}}{\text{capital cost of investment}} \times 100$$

However, it is important to remember that not all benefits can be expressed in tangible terms of costs. Other benefits may include:

- improved job satisfaction for those involved with the new technology,
- improved customer relations,
- improved job delivery times.

COSTING METHODS

The purpose of costing and management accounts is to enable the financial data relating to specific areas and production operations to be evaluated. It also isolates and analyses variables and controls within these areas.

There are three main methods of costing:

- *Absorption costing*: In this, the main objective is to ensure that all the costs are distributed to, and absorbed by, the final printed work sold to a customer. Not only are direct costs included, but also a proportion of the overheads costs, for example of administration and buildings.
- *Standard costing*: This method of costing is based on hourly rates for production operations. Standard production norms or previous experience of costs for different types of work can be used. This method of costing is based on actual records of production, and it is therefore vital in this method that all the records are as accurate as possible.
- *Marginal costing*: The marginal costs can be defined as the variable costs incurred during the production of a single unit of product or output. Marginal costing is a system in which only variable costs are related to the printed product, and it is based on the assumption that production must incur certain costs before any profit is realized. The method recognizes the variable costs involved in print production by individual job, through addition to the prime

cost and variable overheads. Fixed costs are not allocated to work, but are entered into company accounts at various intervals. An important aspect of marginal costing is the contribution margin. The contribution made by a product is its sales price less any variable costs associated with it. Thus the margin represents the contribution that the printed product makes to the fixed costs and profits. Marginal costing is commonly used in the printing industry, especially for decision-making purposes

Breakeven point

Breakeven is the relationship that exists between volume, costs, contribution and profit, and it is the point at which the sales income equals the total costs incurred. It is a situation in which there are neither profits nor losses. A breakeven point can be calculated as follows:

$$\frac{\text{fixed overhead}}{\text{contribution to sales ratio}} = \text{breakeven sales value}$$

Added value

Added value is the income that remains after the costs of materials and other services, such as subcontracted work, have been deducted. The added value is thus the final selling price of the work minus the costs of materials and outwork. The added value is shared between the various financial centres, and used to cover, for example, fixed expenses, wages and salaries, and pension schemes.

All members of staff can contribute to increasing added value by adopting efficient production methods, and achieving optimum production times with reduced wastage. The keys are improved management techniques, and a new approach to marketing in which the markets that are most suited to the printer's technology and the skills of the staff are developed.

The following equation is used to set an annual added-value budget:

$$\frac{\text{salaries and wages} + \text{overheads} + \text{profits}}{\text{total hours available in each cost centre}} = \text{added value in £/h}$$

BUDGETS

Although it is possible to manage the company's financial affairs by responding to events as they arise, this method of operation is very unwise. All printing companies should operate some form of financial control in the form of budgets. Budgets are intelligent forecasts or targets for expenditure in a number of accounting areas, and they are necessary for effective company planning. The main purpose of a budget is to present financial information in a way that assists the management to formulate financial policies and objectives. Budgets thus:

- enable working capital to be effectively used and distributed,
- enable management to make special decisions and prepare long-term plans relating to technology, premises etc.,
- enable human resources to be used efficiently,
- allow the efficient use of technology to be planned, and production methods to be used correctly,
- allow management control of finance in current operations and future short-term operations,
- assist in the prevention of waste in all areas of the company,
- allow the delegation of financial control to various sections and departments,
- provide a standard by which the performance of all areas and departments can be measured,
- reveal any important financial variances.

Some of the areas for which budgets may be prepared are as follows:

- *Sales budget*: This is a budget for investment in market research and analysis techniques.
- *Materials budget*: This allows the planning of finance to be invested in materials to meet production requirements, and it entails consideration of stock depletion or buildup.
- *Labour budget*: This is a forecast of costs and limits to be imposed on various grades and numbers of staff. It is usually derived from the management budget.
- *Production budget*: The production budget is usually linked to the sales budget. It may raise issues such as the optimum use of the existing technology, investment in new technology, the introduction of new methods of working, and subcontracting.
- *Manpower budget*: This entails forecasting the number of staff that will be employed, directly and indirectly. It plans for recruitment where necessary, and any training that is required.
- *Plant-utilization budget*: This identifies the available loads for each area and department. A comparison with the production budget will reveal whether the technology is likely to be overloaded or underloaded.

Budgets may be prepared for any area, as required by the management. They often include management and cash budgets. Special budgets may be prepared for areas and departments whose financial expenditure is felt by the management to be out of proportion. Figure 3.5 shows an example of a company's budget, its actual expenditure, and the variances. Figure 3.6 shows an example of a company's budget and its actual operating expenses.

	Budget 199 –199	Actual expenditure	Variance
Wages, salaries	174 670	182 858	+8 188
Premises	25 838	25 838	0
Furniture and equipment	1 166	1 088	–78
Renewals	820	820	0
Telephone calls	1 988	2 758	+770
Telephone rental	5 198	5 958	+760
Travel and subsistence	1 166	1 052	–114
Advertising	800	1 504	+704
Leased cars	4 434	5 798	+1 364
Interest on stock	23 334	26 002	+2 668
Hospitality	364	198	–166
Training	1 166	630	–536
Total	240 944	254 504	+13 560

Figure 3.5 Comparison of actual expenses with budget

RATIO ANALYSIS

Information about the financial performance of a printing company can be obtained through ratio analysis. This method evaluates the company's accounts against those of previous years, company budgets, and the accounts of other companies in the same field of production.

The calculation of ratios is an important operation as it identifies the financial strengths and weaknesses of a company. It also highlights the performances of the divisions within the company. Many financial indicators can be calculated by ratio analysis. Some of the most important equations used in ratio analysis are as follows:

	Budget 199 –199	Actual expenditure	Variance
Wages and salaries	838 552	855 640	17 088
Operation premises	192 706	181 038	11 668
Hired transport	7 000	1 476	5 524
Central hire charges	4 854	5 756	902
Central hire — fuel	700	1 828	1 128
Totals	12 554	9 060	3 494
Operation equipment	103 834	219 982	116 148
Furniture and fittings	3 500	1 904	1 596
Renewals	274 166	274 166	0
Consumables	1 750	5 598	3 848
Marketing	2 334	6 800	4 466
Insurance	1 166	1 166	0
Books	116	0	116
Other expenses	292	224	68
Travelling and subsistence	584	576	8
Advertising	4 108	0	4 108
Training	2 334	4 042	1 708
Staff clothing	876	2 028	1 152
Stationery	4 394	19 652	13 818
Rectification	2 334	0	2 334
Totals	401 788	536 138	132 910
Gross total expenditure	1 447 040	1 581 876	134 836
Income	584	5 098	
Total income	584	5 098	4 514
Net total expenditure	1 446 456	1 576 778	13 032

Figure 3.6 Comparison of actual operating expenses with budget

$$\text{net percentage profit} = \frac{\text{net profit}}{\text{net sales}} \times 100$$

$$\text{liquidity ratio} = \frac{\text{current assets} - \text{stock} + \text{work in progress}}{\text{current liabilities}}$$

$$\text{percentage return on capital invested} = \frac{\text{profit}}{\text{capital invested}} \times 100$$

ESTIMATING

It is not possible to charge a customer a fair price for the production of printed work unless the true costs of production are known. True costs are calculated using efficient costing and estimating methods. Accuracy in costing is an essential basis of the estimating process. Should the estimated prices be too high, work may be lost, and if they are too low, there will be a loss of profit, or possibly no profit at all.

The estimating operation does not require expertise in mathematics, but a facility for figures is helpful. The most important aspect of efficient estimating is a sound knowledge of all the technical methods and processes used, and an ability to visualize the sequence of operations required for the job. The great majority of print work is unique to the customer, and it must therefore be estimated accordingly.

Estimates are usually based on fixed hourly production rates and production norms. The hourly production norm for any operation or item of technology can be based on experience gained in the past.

Estimates are an intelligent forecast of all the costs that are likely to be incurred in producing a job. The degree of intelligence used, and the technical and production knowledge possessed by the estimator, determine the accuracy of the estimates. An intelligent, informed approach is essential so that each job and its production is evaluated, and the opportunity is taken to use the most suitable methods. A well balanced view of estimating must be taken by the estimator. An overoptimistic view of what can be achieved within the company may result not only in financial loss, but also probably in production-worker frustration because of attempts to achieve impossible production times. An opposite approach in which an overgenerous allowance is made for production times may result in their increasing to unnecessarily high levels.

Three areas of cost are taken into account when estimating:

- materials,
- labour,
- expenses.

Any of these may be directly related to a cost centre (i.e. be direct costs) or indirectly related (indirect costs). General specifications to be considered when preparing an estimate are:

- the nature of the artwork (e.g. the numbers of transparencies, colour prints and monochrome illustrations),
- the volume of DTP,
- the job description (book, brochure, stationery, poster etc.),
- the page format (A4, A5, landscape or portrait),
- the page extent (e.g. the number of pages, the amount of colour),
- the substrate (paper, card, board etc. weight),
- the imposition and the number of colours, including any Pantone colours, or other self colours (assess for any printing difficulties),
- the specification for the covers and any inserts,
- the finishing or binding type (saddle stitch, side stitch, perfect binding etc., lamination, foil blocking, coating etc.).

Questions about printing and finishing technology must also be resolved, as follows:

- What is the number of colours that any one press will print at one part of the sheet, e.g. one, two, four or six colours?

- Does the press perfect?
- Are any coating facilities available?
- What is the maximum sheet size of the printing press and folding machines?
- What are the average production speeds for the chosen technology.

The answers to these questions are then combined with the general job information, and the following issues are considered:

- the number of pages to be printed at view, e.g. eight, 16, 32, and whether they are to be printed half-sheet, or sheet-work. (Consider whether any compound or combined schemes or imposition can be used),
- the amount of scanning or camera work required, and the number of plates to be used,
- the quantity of paper or other substrate required and the size to be ordered,
- the quantity of plates, film, inks, and other special materials required,
- the makeready or setting-up times for all production processes,
- the binding, finishing or other processes and the materials required,
- the delivery requirements, e.g. the methods of packing and despatch.

Information is then required about:

- the costs of inks and coating materials,
- the costs of paper and other substrates,
- the allowances for waste during production,
- the hourly rates for all operations,
- the average ink usage for various types of work,
- the average production norms for each area of production.

This list is not exhaustive, but it provides an indication of the areas of information that must be available before estimation can be attempted. Accurate and efficient estimating requires orderliness, both physical and mental. Working on odd pieces of paper will create problems in both recording and retrieving information. Obviously, estimating can be carried out by both manual and computer operation. A guide to the systems available is given in Reference 1.

Figures 3.7–3.11 show the five parts of an estimate form for a small-to-medium company. The form has been designed not only to allow efficient estimation, but also to analyse the actual hours and costs against the estimate, and show any variance. This provides an excellent method of determining the accuracy of the estimates, and the efficiencies of the various production areas.

SUMMARY

- The success of a printing company ultimately depends on its financial management and resources.
- Financial duties are wide-ranging, and they require particular skills.
- Certain financial information is required by law.
- Profit and loss accounts and balance sheets must be provided to all directors, shareholders etc.
- The amount of capital that is available as working capital is very important.
- The pricing of work and other operations is difficult for printers.
- Accounting procedures have a direct effect on the success of the company.
- In the UK, a number of records are required by the Inland Revenue and HM Customs and Excise.
- An analysis of the sources of costs is essential to maintain financial stability.

ESTIMATE SHEET

Job no.______ Estimate no. _____ Customer account no: _______
Customer details ________________ Contact name ______________

Invoice address ________________ Contact tel. no: _____________

Delivery address ________________ Delivery details _____________

Job specification

Product ____________________ No. of pages ______________

Quantity ____________________ Colours/specials ____________

Format ____
Cover/Colours/Specials ______
Flat size ____________________

Finishing ____

Paper stock text ______________________________________

Paper stock cover/other _______________________________

Outwork ____________________ Invoice no. and date ________
VAT ____________________

Figure 3.7 Estimate form. Part 1: job details

Cost centre	Estimated		Actual		Variance
	Time	Cost	Time	Cost	Cost
Camera/scan					
Process					
Screen					
Line					
Tints					
Line shots					
Planning					
Separation/masking					
Contacts					
Film assembly					
Plate preparation					
Number of plates					
Separation at plate					
Printing					
Press					
Makeready					
Run-on					
Wash up					
Ancillary					
Folding					
Statue					
MBO					
Finishing					
Guillotine					
Müller					
Bench Work					
Packing					
Other					
Cost centres subtotal					

Figure 3.8 Estimate form. Part 2: cost centre time–cost breakdown

Materials	Estimated		Actual		Variance
	Quantity	Cost	Quantity	Cost	Cost
Paper, stock, text					
Overs/waste					
Cost per 1000					
Paper/cast stock-covers					
Overs/waste					
Cost per 100/1000					
Negative film					
Duplicate film					
Contact film					
PMT materials					
Masking film					
Plates, type and size					
Inks					
Black					
Process					
Other					
Materials subtotal					

Figure 3.9 Estimate form. Part 3: materials quantity–cost breakdown

Outwork	Estimated price	Quotation from supplier	Actual cost (from invoice)
Outwork Subtotal			

Figure 3.10 Estimate form. Part 4: outwork breakdown

Department	Materials	Labour	Outwork
Studio			
Printing			
Finishing			
Totals			

Figure 3.11 Estimate form. Part 5: departmental analysis of costs

- All cost centres should be segregated for analysis and evaluated individually.
- Depreciation is a very important issue for printers, who normally invest large sums of money in technology etc.
- A number of sources of finance exist. Some of these are very expensive.
- A number of methods are used in costing, each using a different approach.
- Budgets are a necessity for efficient financial control.
- Ratio analysis is used to determine certain financial statistics.
- Estimating requires technical knowledge and an understanding of the best production methods available within the company.
- Estimated costs and the actual costs incurred should always be compared.
- Statistics for estimation can be based on the actual times recorded for similar work, provided that the company is operating efficiently.

ACTIVITIES

- *Activity 3.1*: Identify the information that you consider should be presented in a company balance sheet and profit and loss account.
- *Activity 3.2*: Identify two methods of analysing and apportioning costs for recovery.
- *Activity 3.3*: Outline a method of calculating depreciation.
- *Activity 3.4*: Identify a financial issue that is not dealt with in this chapter.

REFERENCE

1 **Birkenshaw, J W** *Pira Guide to Management Information Systems for Printers* Pira International, UK (1991)

4
Human resources

HUMAN-RESOURCES MANAGEMENT FOR PRINTING

> No matter how much work a person can do, no matter how engaging their personality may be, they will not advance far in business if they cannot work through others. (*John Craig*)

To printers, the term 'resources' often means the technology and materials they use in production. However, technological resources are of little significance unless they are combined with the human factor, as people are the lifeblood of any organization. The printing industry, in common with other industries, is experiencing rapid changes in both technology and methods of production. This, in turn, results in a major change in demand for human-resources provision. Increasingly, it is recognized that any printing organization's reputation is allied to production efficiency, and this depends to a large extent on the quality of its staff. Quality is defined in this context as the correct attitude, ability and experience, and, most importantly, experience must be relevant to the needs of the company. One company's needs are not necessarily those of another engaged in the same production process. With this in mind, we need to recognize that an individual will, through personality and attitude, develop towards producing a particular type and quality of work, and one who is used to the demands of high quality and detailed production will not normally be happy carrying out work for which the prime consideration is price, and the resultant lowering of quality. For many production operations, genuine experience is necessary. However, the relevance of the experience being offered to the company by an individual requires analysis, an example being the machine printer who has previously been engaged on multicolour production, and has over 20 years experience of the manual production control of quality. This experience, although valuable, would not be relevant to the requirements of electronically controlled multicolour printing without training and induction. Also, while the advent of desktop publishing has opened the composition (wordprocessing) operations to all, the experience gained by a professionally trained compositor will certainly allow him/her to carry out wordprocessing and page makeup to a higher standard than the individual with less experience. Throughout the printing industries in the UK and continental Europe, fewer people are employed in print production than were a few years ago. This is not the result of trading conditions alone; rapidly advancing technology leads to the realization that fewer people can, by adopting new production methods, achieve considerably more than was previously achieved by greater numbers of staff. The situation is not likely to change in the foreseeable future. However, a negative attitude should not be adopted, but rather a positive view that capacity will increase and returns on investment improve, under normal trading conditions.

Also, the imposed requirement in the past on printing management to work and communicate through the representatives of collective groups of employees has, to a large extent, diminished; it is now increasingly possible to communicate with staff as individuals. Rigid demarcation has given way to sensible levels of flexibility within areas of production specialization. An outstanding example of this situation is seen within the origination area of the industry. Origination, with the increased use of electronically generated production methods, has evolved to cover all production operations, from initial wordprocessing through to printing-plate production. In press production, ink-on-paper operations, a decline in staff, machine-type specialists, is seen. A preference has developed for an overall ability and understanding of printing techniques and the electronic control of production. Labour-intensive operations in print finishing have given way to automation. These ongoing changes provide an excellent opportunity to review human-resources provision for the printing industry in a way previously impossible. Obviously, the skills required in the various areas of production are today very different from those conventionally used for many years. Some skilled operators may well adapt to new methods successfully, while others, who one would think would quickly adapt, owing to previous experience, fail. This creates a situation in which a school or college leaver may be more successful than a well established, mature and experienced person.

Allied to the rapid advance in technology has been the widespread fear within the printing industry that technology is about to replace people. Quite naturally, in times of change, people feel threatened by new methods of work and equipment. Why should fewer people be employed in printing than a few years ago? Undoubtedly, the major influence has been an increase in understanding by both employers and employees that achieving and maintaining competitive ability is essential to survival. Realistically, we cannot do today's work with last year's methods and remain in business in the future. When a technology and human-resources situation requires resolution, it should not be viewed in a negative manner by the company. Any possible redundancy, retraining or redeployment should not be viewed as the fault of new technology alone. Changes in work demand together with price and quality requirement play their part, resulting in the company having to assess its position and future strategy. Termination of employment should only occur in the absence of any alternative. Dismissal nearly always has traumatic effects, not only on the individual concerned, but also on other employees.

Even when the dismissal is fully justified, staff sympathy for the individual often exists. If more than one member of staff is involved, a general feeling of 'who is next' develops.

The retraining or redeployment of staff affected should be the first consideration. Providing all the circumstances are clearly communicated to staff, cooperation and transfers to other duties may be possible. If not, early retirement or voluntary redundancy may be the kindest way to resolve the situation. It is with these important facts in mind that we must consider a crucial factor: people as individuals are probably the most important resource of any organization, and until people are well trained, educated, well motivated and well managed, can they provide a successful contribution towards their company's stability and development? The efficient management of human resources therefore becomes more important to the health and status of a company than was ever the case in the past. The main functions of human-resources management are shown in Figure 4.1.

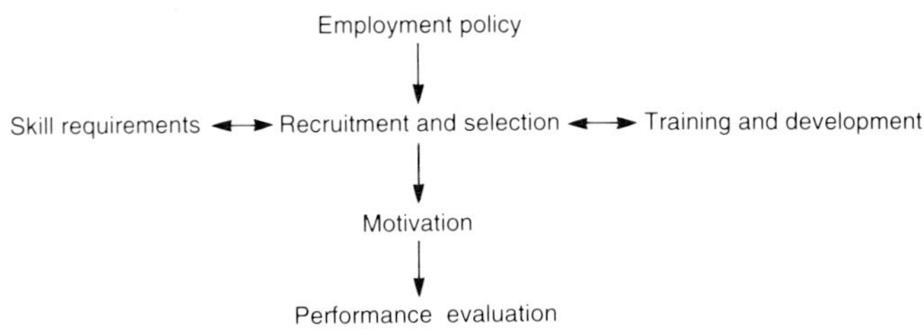

Figure 4.1 Functions of human-resources management

EMPLOYMENT POLICY

Successful printing companies of all sizes pay close attention to employment policies and practice. An efficient employment policy states a company's thoughts and intentions on a wide range of issues relating to human-resources provision and management, for example the following:

- work organization, environment and ethos,
- recruitment, selection and evaluation,
- pay and conditions of employment,
- security of employment,
- resource planning,
- training,
- communications,
- training and development,
- personnel records.

A carefully and clearly constructed employment policy can assist in the reaching of decisions related to staff management. Decisions are generally understood and applied evenly and fairly throughout the company. Also, the implications of decisions on other areas of the company are apparent and, most importantly, satisfy legal requirements.

The management of human resources is easier and more effective if a clear indication is provided of how the company's objectives can be achieved and procedures put into practice. It can also remove the demotivating effect on staff of what appears to be any management uncertainty and inconsistency in decision making.

An excellent UK guide to developing an employment policy is contained in the Advisory, Conciliation and Arbitration Service (ACAS) Document 10. This states that the preparation of employment policies needs to take account of influences from both inside and outside the organization, for example:

- legislation or legal decisions that require the development of particular policies (for example the *Health and Safety at Work Act* applies to all work activities in the UK and it places a positive emphasis on the need for the joint involve-

ment of employers and employees in the development of health and safety policies and efforts to reduce risks),
- existing policies in other areas of company operations, for example that on quality control, in relation to recruitment and training policies,
- company agreements that affect a particular area,
- prevailing good practice (for example, a policy on the harmonization of the terms and conditions of a company's administrative and production workers could be influenced by corresponding developments in comparable companies).

No one set of policies or procedures is suitable for all companies, and they will fail if they are not carried out by understanding and effective managers. The key issues must be developed to meet the needs of the individual company. It is in this context that the views of line managers and other employees can be taken into account. Precise job specifications that are entirely suitable for a large printing company may well be completely inappropriate for a small printer who requires total flexibility of staff in production operations. Policies relating to quality control for the printer producing high-quality multicolour work etc. may be both unnecessary and financially impossible for the general and commercial printer engaged in average day-to-day print requirements.

It can be seen that employment policies deal with a company's aims, intentions and structure in relation to human resources.

Managers will be more effective if they are clear about the company's objectives; employees can also be more effective and gain greater job satisfaction.

EVALUATION OF SKILLS REQUIREMENTS FOR PRINTING OCCUPATIONS

In a commercial market characterized by strong competition and an ever-increasing complexity of technological advances, attention must more than ever be directed towards recruiting the right person for the job. In the past, no formal educational qualifications were generally required for entrants to the printing industry, although many entrants subsequently gained trade qualifications at college. However a trend has developed in which companies employ staff for key positions who are college graduates, rather than having come directly from school.

If we accept that the quality of the company's workforce, in terms of ability and attitude, is crucial to achieving success, we have to ask how one recruits the right person for the job, and what the criteria for developing effective human resources in the modern printing industry are. To answer this question, it is necessary to understand the qualities and skills required by employers for the various main production and administration areas. These areas are as follows:

Production:

- design/studio,
- origination:
  - DTP,
  - scanning,
  - planning and print-surface preparation,
  - artwork preparation,
- printing:
  - sheet-fed,
  - web-fed,
- finishing:
  - auto/mechanical,
  - hand/bench,
- bookbinding:
  - hand
  - mechanical

Table 4.1 Production-worker practical skills requirement (essential skills)

		Creative	Science	Technology	Electronics	Mechanical
Origination	DTP	xx		xx	x	
	Scan	x	xx	xxx	xx	
	Plan/surface preparation		x	x	x	x
	Artwork	xx				
Printing	Sheet		xx	xxx	xx	x
	Web		xx	xxx	xx	xx
Finishing	Auto/mechanical		x	xx	x	xx
	Hand/bench	xx	x			
Packaging	General	xx	x	xxx	xx	xx
Binding	Hand	xxx	xx			
	Mechanical		xx	xx	x	x

Creative: innovation, development of new ideas, ability with design skills; *Science*: knowledge of science as applied to production processes, materials etc.; *Technology*: knowledge of technology in use for the various production operations; *Electronics*: knowledge and ability as applied to the control of quality and production; *Mechanical*: design of technology, configurations, ergonomics etc.
x: basic knowledge and ability, xx: working knowledge, xxx: in-depth knowledge.

Essential skills are those required by the job, and not those gained while doing the job if training is offered.

Table 4.2 Administrative worker skills requirements (essential skills)

	Creative	Finance	Technology	Control	Policy	Liaison	Communication
Section manager/supervisor	x	x	xx	xxx	x	xxx	xxx
Department manager	xx	xx	xx	xxx	xx	xxx	xxx
Sales/marketing manager	xxx	xxx	x	xx	x	xxx	xxx
Finance/estimating/costing		xxx	x	x	x	xx	xxx
Managing director	xxx	xxx	x	xx	xxx	xx	xx
Secretary/PA		x		x	x	xxx	xxx
Design	xxx	xx	x	x		xxx	xxx

Creative: innovation, development of new ideas, ability with design skills; *Finance:* knowledge of business and production financial issues; *Technology*: knowledge of technology in use for various production operations; *Control*: organizational strategy and management abilities; *Policy*: Aims and objectives of company; *Liaison*: clients and interdepartmental; *Communication*: effective communication of all issues affecting company.

x: basic knowledge and ability, xx: working knowledge, xxx: in-depth knowledge.

Essential skills are those required by the job, and not those gained while doing the job of training is offered.

Administration:

- section manager/supervisor,
- department manager,
- works production manager,
- sales/marketing manager,
- estimating/costing/financial,
- secretary/PA,
- managing director.

When a company decides to recruit a new member of staff, certain questions should be asked:

- What exactly is the vacancy that is to be filled?
- Is it the same job as it was when the previous holder carried it out?
- What existing skills are required for the job to be successfully carried out (see, for example, Tables 4.1 and 4.2)?
- Are these skills likely to continue to be required, or will they change in the near future?

The importance of good employee selection and recruitment procedures cannot be overemphasized, as employing the wrong staff will have a direct and unfortunate impact on the organization's performance in the short term and the long term. A well thought out job description and specification linked to correct selection will help to raise poor performances to high levels of production in all areas.

High organizational performance, and consequently high human performance, is becoming imperative if organizations are to remain in business in a complex and highly competitive industry. Thus the critical process of finding and keeping the right employee assumes great importance. The difficulty of recruiting to a position increases in direct proportion to the importance of the position, and therefore it is necessary to ask whether it is necessary to make an external appointment. Is it possible to make an internal appointment, and then make an external appointment for a lower position? However, many companies prefer to appoint externally, with the idea of introducing 'new blood' and therefore new thinking.

Can the workload be redistributed or reduced by mechanization? The thought processes involved in employee recruitment and selection provide an opportunity to assess the situation and rethink towards change. The process of recruitment begins with the realization and confirmation that a position exists and needs to be filled, and continues with reference to business plans, through human-resources budgets, the completion of relevant job descriptions and specifications, to interview, appointment and induction.

The analysis of labour requirements is essential. Although these requirements can only be forecast with any accuracy in the short term, a long-term forecast must be attempted if the company is to have any chance of controlling its future. Analysis should concentrate on:

- the structure and number of staff analysed by department or skill, together with any plans for expansion or contraction, centralization etc.,

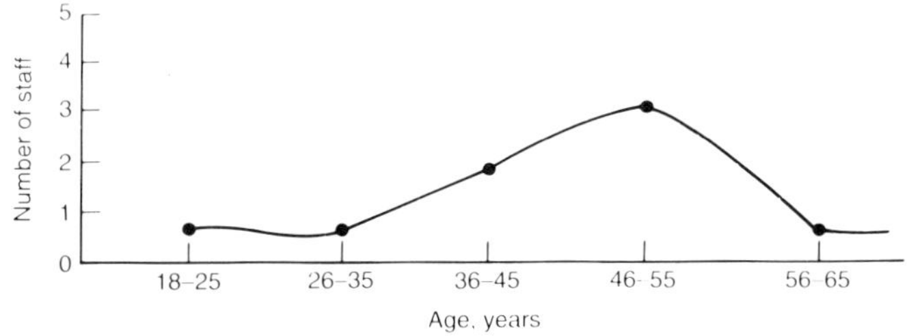

Figure 4.2 Staff age profile

- the age profiles of staff analysed by department and skill to determine succession trends and requirements (see, for example, Figure 4.2),
- the impact of any new proposed technology on manning levels,
- the development of individuals within the company, e.g. whether existing staff are promoted or external staff are recruited,
- the company aims and objectives,
- the total labour costs.

Staff-resources planning can be carried out using management judgment, present and expected future situations within the company, and/or a review of statistical data to reveal trends and requirements. Data may also be obtained through work-

Table 4.3 Staff data form

Department: Printing Date:

Name	Length of service, years	Age	Duties	Comments
H Evans	4	31	Small offset printer	Average ability and skill
B Jones	16	44	Small offset printer	Handles majority of class work
J Smith	30	51	Letterpress plater	Specific skill, demand now in decline
S Clark	10	49	Two-colour A2 press	Capable of high quality production
M Black	6	49	General assistant	Good assistant, no ambition to progress
J Potter	25	60	Single colour A2 press	Works fast, quality can be variable
S Andrews	2.5	24	4-colour A1 press	Excellent printer, conversant with electronically controlled production; acts as No 1 printer
A Brown	8	45	4-colour A2 press	Average ability, acts as both No 1 and No 2 printer

study techniques. The preparation of a staff-planning chart such as that shown in Table 4.3 assists in planning.

Job description and specification

Analysing the job by preparing a job description and specification is recommended. The following points should be considered:

- *job description*: a broad statement of the scope, duties and responsibilities of a particular job,
- *job analysis*: the process of examining a job to identify its component parts and the circumstances in which it is performed,
- *job specification*: a product of the job analysis that is a detailed statement of the mental and physical activities entailed in the job (the specification is usually expressed in terms of behaviour, in other words, what the worker does, what knowledge he/she uses in doing it, and the judgments made and the factors taken into account when it is done),
- *personnel specification*: an interpretation of the job specification in terms of the type of person who is suitable for the job.

Employment considerations

Employment considerations can include some or all of the following:

- interesting work,
- opportunities for training and development,
- competitive rates of pay,
- incentive schemes (merit, productivity, and profit sharing),
- pension schemes,
- payments in kind (car, health benefits etc.),
- working hours and conditions,
- leave entitlements, sick pay etc..

It is important to survey the selection process and appraisal systems as mistakes in selection, even in the lower grades, can be very expensive to rectify.

The benefits derived from analytical selection include:

- increased employee satisfaction,
- less waste and spoilage,
- lower absenteeism and labour turnover,
- improved methods and systems,
- an increased level of output,
- a smaller supervisory burden,
- fewer grievances,
- improved morale,
- greater cooperation.

Example job description
Job title: Production Manager

Reporting to: Managing Director

Responsibility: The organization and maintenance of production within the company to meet efficient productivity and client demands.

Duties:

- To receive all work for production and to assess the most efficient production methods.
- To schedule work in relation to the client delivery date and the optimum use of technology within the company.
- To oversee the preparation of work technical specifications.
- To organize and control the staff involved in production.
- To provide necessary production data to senior management.

It is important to ensure that:

- the description conveys a clear and accurate overall impression of the job,
- the description is of the *job* and not the person,
- the description does not contain abbreviations or ambiguities.

Selection criteria

- *achievement*: general education, examination success, progress since leaving school or college,
- *special abilities*: industrial examinations and qualifications, subject specialization, experience gained and where,
- *disposition*: ambition, perception of self, aggression, calmness etc.,
- *physical attributes*: general health, time off sick, ability to meet required physical demands,
- *personal circumstances*: name, address, age, marital status, any problems (travel etc.),
- *interests*: are any of these of occupational significance?

Essential skill requirements

The essential skills requirements are those that are critical if the job is to be performed to satisfactory standards.

Desirable attributes

The desirable attributes are the less essential characteristics, which may be little more than preferences.

Contraindications

The contraindications are those attributes that disqualify an individual or group of individuals even if all the above are satisfactory.

RECRUITMENT

Selection aims

- Have a thorough understanding of the job being selected for, that is, to analyse the job.
- Be quite clear about the kind of person that it is necessary to attract to fill the job.
- Be aware of the strengths and limitations of the possible ways in which a person can be assessed for the job.
- Be aware of the importance of the careful introduction and later follow-up of the person chosen for the job.

Important points to consider when recruiting

- Avoid vague statements, and make the description of the person as specific as possible.
- Make sure that a job description is prepared.
- Be clear about the educational and intelligence levels required for the job. Too low a level will lead to poor performance, and too high a level will lead to dissatisfaction and labour turnover.
- Remember that the purpose of recruitment is to fill a position, and not to advertise the company.
- Be specific about salary and working hours.
- Ensure that details and actions are within labour law.

Interviewing

The interview should be conducted by someone who is skilled in the process, and it should follow a carefully thought-out structure and planned technique. Unfortunately, this is often not the case, the interview being a short period of discussion in which some of the key issues are not clearly dealt with. Unfortunately, the majority of printing managers are not good at interviewing, and, although they are quite aware of the importance of the interview for both the applicant and the company, they still do not prepare in advance. It cannot be overemphasized that the interview plays a crucial role in achieving the company's success; a wrong appointment can be very difficult to remedy. Some points that assist in successful interviewing are the following:

- Maintain a friendly and open manner throughout the interview.
- Prepare any questions to be asked in advance.
- Refrain from stating personal opinions and making personal statements.
- Ensure privacy, and avoid interruptions.
- Do not make promises that may not or cannot be fulfilled to secure a suitable applicant.
- Analyse interviewees' skills and abilities in relation to what is required in the situation being interviewed for. Remember that many applicants may have extensive knowledge and skills, but not necessarily in relation to the skills required in the position.
- Make it clear whether any training is being offered.
- Make sure that all the documents that will be required during the interview are in place. An interviewer who pauses to locate misplaced papers etc. loses control of the interview, and promotes an impression of inefficient management.
- Make the salary level, hours of work, holiday entitlement etc. clear.
- Do not become emotionally involved.
- Get the applicant to talk; remember that it is an interview.

JOB EVALUATION

The contribution that each individual job makes to the overall success of a company requires periodic evaluation. Job evaluation and employee performance appraisal are important parts of a printing manager's work, and they are encompassed within the overall management of human resources. The past decade has seen a dramatic shift within the printing industry towards a more formal approach to performance evaluation. In simple terms, the contribution of each job needs to be planned, implemented and then evaluated. The results of correctly organized evaluation include the following:

- the identification of an individual's ability to meet the job requirements,
- the identification of promotion potential,
- the provision of a basis for the reward or encouragement for individual effort,
- the identification of training needs,
- the improvement of communications relating to individual job requirements between different levels of authority,
- assistance in future human-resources planning and strategy.

Evaluation can be classified into factors, as follows.

Factor 1: Complexity

- assessment of the complexity of the tasks entailed in doing the job, and measurement of the level of thought and the frequency of thought required to achieve the job objectives,
- the main purpose of the job and the areas of responsibility,
- predictable or unpredictable situations or pressures that may arise during the course of the job.

Factor 2: Human relationships

The human-relationships factor evaluates the extent of the human relationships relating to the job, as follows (the emphasis is on purpose, reason and contact, and not the level or the status of the job):

- *Range*: The range and level of contact with other staff members and clients must be evaluated.
- *Depth*: The degree of contact in relation to the provision of information, team working, the motivation and supervision of staff etc. must be measured.
- *Contacts*: The main working contacts must be identified.

Factor 3: Initiative and judgment

Factor 3 evaluates the extent and level to which initiative and/or judgment needs to be exercised. It takes into account the responsibility of the job holder for the planning and control of the workload of the job holder and others. The responsibility for expensive technology or production materials is assessed.

Factor 4: Levels of responsibility

Factor 4 evaluates the level of responsibility within each job for the company's assets, both quantifiable (finance, staff and technology) and unquantifiable (organization/company reputation and prestige). The evaluation of the responsibility for errors of the job holder is critical, and errors are classified into four groups:

- errors that are containable, and for which any effect on quality and the company can be easily and quickly rectified,
- errors that interfere with the day-to-day operation of the company, and take time to correct,
- errors that disrupt an area of the company, and have implications for other areas, and take time to rectify,
- errors that cause considerable damage to the organization/company's reputation, and lose clients and production, and for which considerable expertise is required to rectify the situation.

Factor 5: Essential skills

Factor 5 evaluates the minimum amount of life and work experience, training and practical skills that are essential to enable the job holder to carry out the job to the standard required. Essential skills should be regarded as those that are initially required by the job, and those gained while doing the job.

Education (level of education required to be able to do the job):

- education to school-leaving age,
- education to O/A-level standard or equivalent,
- education to degree level or equivalent.

Training and skills (extent of specific or specialized training requirements):

- up to one year,
- 1–2 years,
- 2–3 years,
- three or more years.

Factor 6: Physical requirements

Factor 6 evaluates the level and frequency of physical effort and activity required to do the job, as follows:

- limited, e.g. sedentary, standing,
- moderate, e.g. standing, continuous movement, carrying,
- strenuous, e.g. heavy lifting.

Frequency:

- minimal (up to 20% of working time),
- frequent (20–70% of working time),
- continuous (over 70% of working time).

Factor 7: Special conditions

Factor 7 evaluates any unusual and demanding elements of the job. The conditions must be an integral part of the job that are not rewarded by any special allowances or increments. Examples are as follows:

- *physical environment*: unpleasant physical conditions, such as excessive noise, fumes, dirt and restricted working space,
- *mental stress*: stress caused by conditions such as working in isolation, or monotonous repetitive work,
- *social factors and conditions*: for example, regular unsociable hours, inconvenient extensive travel.

STAFF APPRAISAL

Example staff-appraisal forms

An example staff-appraisal form for production staff is shown in Figure 4.3, and a form for management staff is shown in Figure 4.4.

MOTIVATION

> The person who gets the most satisfactory results is not always the one with the most brilliant single mind, but rather the person who can best cooperate the brains and talents of associates. (*W Alton Jones*)

What makes an employee want to work? Every manager is aware of how complicated individual motivation can be. In this context, aims throughout the company must be directed towards greater awareness and acceptance of the importance of achieving an improved competitive position. Industrial psychology applied to human-resources management and motivation is a valuable instrument for dealing with human motiva-

STAFF-APPRAISAL FORM: PRODUCTION STAFF

Name ______________________________
Job title ______________________________
Job grade or status ______________ Location ______________
Time in this position ______________ years ______________ months

	5	4	3	2	1
Quality of work (Accuracy to standard, e.g. achievement of technical work specifications and client requirements. Maintenance of quality.)					
Quantity of work (Quantity of work acceptable under normal conditions, e.g. percentage of production in relation to accepted norm.)					
Knowledge of job (Clear understanding of the technology, technical operations, methods, work and equipment involved.)					
Cooperation (Attitude to, and willingness to work with, supervisors to achieve results.)					
Dependability (Conscientious with respect to punctuality and attendance. Loyalty to company.)					
Initiative (Able to proceed alone without close or constant supervision.)					
Health (Absence from work due to health problems.)					

Enter a cross in the appropriate column for each aspect of performance (5 is the highest score and 1 is the lowest — turn the form sideways to see a profile of the member of staff. List any significant changes required below.)

Comments: ______________________________

Signature of appraiser ______________ Date ________
Job title of appraiser ______________________________

I, the appraisee, have read this evaluation, and I (a) agree with it, and/or (b) wish to make the attached comments (delete as applicable).

Signature ______________ Date ________

Figure 4.3 Staff-appraisal form for production staff

STAFF-APPRAISAL FORM: MANAGEMENT STAFF

Name __________
Job title __________
Job grade or status __________ Location __________
Time in this employment __________ years __________ months

Name of appraiser __________
Job title of appraiser __________

Significant changes within the company since the last review:

Achievement of the job holder against the management objectives set out in the company business plan and the holder's job description:

List any particular aspects of the job that were performed particularly well.

List any aspects of the job that were not performed particularly well or that indicated cause for concern.

List any self-development achieved or training courses attended by the manager.

Indicate the manager's level of cooperation with other managers and senior managers.

Rate the manager's performance over the appraisal period.

Additional comments on performance rating.

Figure 4.4 Staff-appraisal form for management staff

What has the manager achieved during the appraisal period?

Main objectives	Results	Rating

Review the personal development of the manager in relation to the results achieved.

Management skills, planning, organizing, delegation and control (consider strengths and weaknesses):

Personal qualities (list any personal features that have an implication on the manager's job performance):

Attendance and punctuality (state reasons for any problems):

Further training and development required over the next appraisal period:

Appraiser's summary:

Signature of appraiser ______________________________ Date _________

Comments of appraiser's line manager:

Signature of line manager ______________________________ Date _________

I, the appraisee, have read this evaluation, and I (a) agree with it, and/or (b) wish to make the attached observations (delete as applicable).

Signature of appraisee ______________________________ Date _________

Appraisee's objectives for the next appraisal period:

Figure 4.4 continued

tion in both production and administration situations. This is very important, as:

- achieving a high quality of workforce and the correct attitudes of workers is crucial to success,
- the culture of initiative in the workforce is preferable to mere obedience, as proficiency, and therefore productivity are not dependent on skills alone, but also on the utilization of an individual's abilities and energy.

Care must be taken to differentiate between motivation and incentive. Motivation can be encouraged by external actions, but it must also spring from within the individual, and may well remain in force if the external action is withdrawn. Incentives are forces applied externally to encourage achievement; however, the effects of incentives cease as soon as they are withdrawn.

Motivating staff has long been a major problem for managers. Obviously the discerning manager will motivate staff as much as possible, as the result is proved to be a higher level of utilization of employees' abilities in a more effective and efficient company, and greater employee job satisfaction.

Examination of A H Maslow's theory of motivation reveals that people are motivated by five basic needs, the primary and basic need being that of physical survival, and self-fulfilment being the highest level of need. These needs are often shown in pyramid form, as in Figure 4.5, and as listed below:

- *Physiological needs*: Physiological needs are those that are basic to continued existence. Air, water, food, rest, money, clothing and shelter are regarded as the basic needs that motivate an individual to work.
- *Safety needs*: Safety needs, such as physical and psychological security, are connected with protection, and they include adequate sustenance and shelter. These having been satisfied, the individual looks for a measure of security and protection against the lowering of his/her living standards.

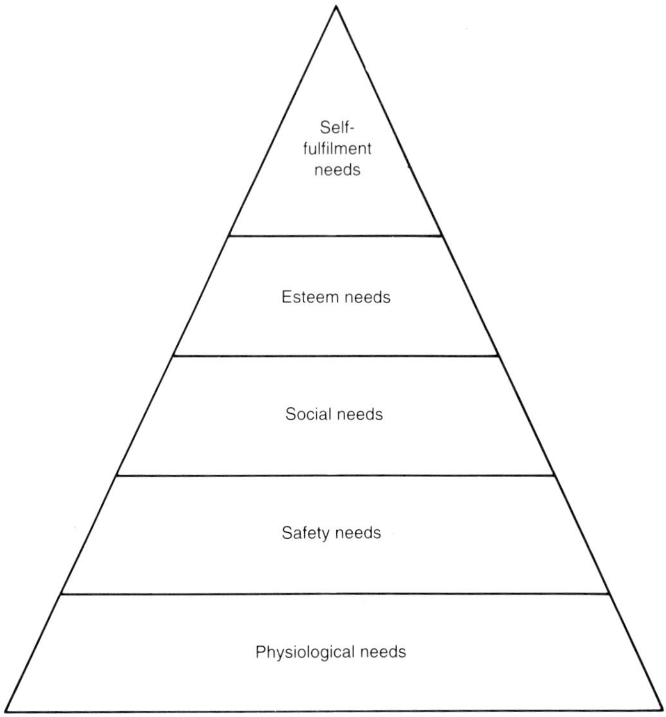

Figure 4.5 Needs hierarchy

In the absence of any threats, the individual feels secure, and is motivated by the next set of needs in the pyramid.

- *Social needs*: Social needs are those which motivate a person to enter actively into a social environment, and they can be identified as love, attention, social communication, acceptance and friendship. In the workplace, these needs are identified as social/work groupings, i.e. trade-union involvement, welfare groups etc., and they are distinguished by the need to be accepted by peers.

- *Esteem needs*: Esteem needs are those which follow from self-satisfaction, i.e. self-respect, independence, prestige, status and recognition. Status gained from the esteem of others is an important factor. With few exceptions, people need and seek the approval of others, and gain status from it. Failure in this area can often be seen in an individual's feeling of inferiority; this always results in a lowering of morale.
- *Self-fulfilment needs*: Self-fulfilment needs, defined as self-actualization needs by Maslow, are at the peak of the pyramid, and they can only exist after the previous four have been realized. Self-realization, self-fulfilment, accomplishment, personal creativity and identity are the identified needs. It is these needs that lead to creativity and the seeking of power through authority.

The great majority of printing-industry workers earn enough to cover their psychological and safety needs, and thus Maslow's higher needs require attention.

Not all motivation is completely financially based; in fact, finance is often not a consideration. Managers must develop the awareness that people do not always work for money alone. Some nonfinancial motivations are as follows:

- *Personal goals and objectives*: Many workers are motivated to achieve objectives of career progression, and are willing to forego any short-term financial advantage. However these individuals must see their way to the top, or they will become demotivated, and seek opportunities elsewhere.
- *Social aspirations*: Many workers have social aspirations such as good educational provision for children, and the membership of professional bodies, social groups and institutions. These social aspirations are nearly always found in individuals who aspire to management.
- *Use of developed skills*: Many workers in the printing industry are proud of their skills, and wish to achieve positions that will allow them to use their skills, and thus obtain job

TRAINING

The costs of training are viewed with doubt by some managers who have not assessed the true costs of incorrect recruitment and little training. The apparent costs of training, however, may hide far greater costs incurred on such issues as:

- unnecessarily high manning levels and overtime payments,
- lost or delayed production,
- interruptions to the flow of work,
- increased production costs, waste of materials, and risks of accidents to staff and machinery,
- low morale, resulting in low productivity,
- damage to the organization's reputation.

It is essential, therefore, to restrict recruitment, training operations and decisions to those who are technically competent, and to emphasize any training that may be required. Often, managers are so immersed in the day-to-day technical problems of operation that training is overlooked, and yet this is the time when it is most needed. The only way to resolve day-to-day and ongoing problems is by updating skills through training and development. Training does not always have to be highly technical, and it can, for example, on some occasions, be facilitated by the establishment of a discussion group in which staff may express and discuss current problems. Subsequently, a quality circle can be formed. These circles meet at regular intervals, and discuss production and quality issues.

The company's true need for employee training and education can be assessed from job descriptions and the results of job evaluation and employee appraisal.

The major thrust of training is on the achievement of manual skills; however, the educational aspect of developing background skills and mental faculties is also very important. The quality of the printing-industry workforce for the future is being determined today by the quality of external and in-company training schemes and objectives. Currently, college and training agencies' requirements are for student competence in a number of specialities in selected areas. The emphasis is on core skills that it is hoped will develop a flexible skilled employee. Fortunately, the demands for narrow skills training, i.e. machinery-operator skills alone, is fast receding, the emphasis being towards the development of transferable skills within broader areas. This is obviously preferable, as the management of technological change is far more easily achieved with this type of trained and educated employee.

Training and education may be achieved within the company, or externally, at university, college or polytechnic. External training can be at all levels, from trainee to graduate.

A trend is developing of printing companies drawing staff from college graduates, the recognized advantage being that the applicant will have received training, and also the knowledge to back training in an industrial setting. From this starting point, a company can enter into a programme of specific skill development that is related to the company's needs, with the advantages of lower training costs and faster employee progress. The great majority of professional training, i.e. management development, is completed in colleges, and is assessed by formal examinations that may result in admittance to professional institutions such as the UK Institute of Printing and the Institute of Management.

A recent development is the provision of a degree course for printing management. Traditionally, printing workers progressed from the shop floor to various management positions. However, a good production worker does not necessarily make a good manager. The failure of this type of progression to management lies in the lack of business knowledge acquired by a production worker during training and further education. For some time, the printing industry has realized that efficient management has to be developed, and that business knowledge is imperative for success. The importance of proper training and education cannot be overemphasized, as the printing industry, in its many complexities, provides an interesting and demanding challenge to the abilities of managers at all levels.

The benefits of training and education for managers are:

- an improved attitude towards the company,
- a better understanding of the company's and the customers' needs,
- greater understanding by the workforce of the difficulties faced by the management,
- more efficient work with less wastage,
- reduced assimilation times,
- greater workforce flexibility,
- improved management/staff relations.

The most valuable manager is one who is training somebody to be a better person than they are. (*Roy C Ingersoll*)

HUMAN RESOURCES FACTORS THAT LIMIT PRODUCTIVITY

Wherever people are assembled for work, it becomes necessary for managers to organize them in a structured manner. The nature and attitude of the organization can have considerable effects on production. Human motivation has so far been examined from management and social viewpoints, but it is now necessary to examine factors that affect employee behaviour in the production operation:

- *Recognition and reward for high performance*: All staff respond to management recognition of their particular abilities and achievements. Make sure that all employees are aware of company achievements and client satisfaction, and not just of any complaints received.
- *Relations between employees*: Every effort should be made to bring staff together as a team. Staff, and therefore the company, benefit from the combined skills of the workforce. Interdepartmental cooperation is essential, and it can be achieved through staff meetings or the formation of a qual-

ity circle, in which problems relating to production or communications can be discussed.

- *Example of behaviour of management*: The management should be able to work with the staff with sympathy and understanding, display self-control, and reach decisions without prejudice. Staff respond to a management's willingness to accept responsibility and accountability for decisions. Management attempts to shift or avoid responsibility quickly demotivate staff and lead to loss of credibility.
- *Conditions in which people are asked to work*: Poor working conditions in terms of environment and rewards can result in increased absenteeism, increased industrial-relations group activity, low quality, and increased accident rates. While an environmentally advantageous location is not available to all printers, good-housekeeping procedures can be. Staff working in clean conditions invariably perform better.
- *Provision and condition of technology*: The realization by staff that the technology provided to achieve production is either out-of-date or simply incapable of meeting the demands placed on it rapidly demotivates staff and leads to stress-related problems.
- *Nature of work being produced*: Simple operations, often carried out by rote, with long periods of repetition, lead to staff apathy and general loss of interest. Attention should be given to automation where possible. Job rotation is another method of reducing the problem. Staff within a team interchange jobs with each other, thus reducing the boredom generated by reduced mental activity.
- *Adequate instructions*: Correct communication of work technical specifications to departments and individuals is critical in preventing human-resources waste and dissatisfaction. For each production operation, there is opportunity for error. In fact, for every one correct way to work, there are many wrong ways. Major staff dissatisfaction and frustration can arise from inadequate instructions and supervision, and this can lead to staff doubts about the management's ability.

- *Rapid changes in production plans and schedules:* Attempts should be made to avoid frequent short-notice changes to work schedules. This is difficult, because of the immediate-demand nature of the printing industry. However, one has only to experience the frustration of staff that results from sudden adjustments to work schedules to realize the adverse effects on both workers and production.
- *Company reluctance to invest in staff training*: Some printing companies regard investment in training its workforce as unnecessary, and others see developing skilled staff as a positive investment. Printing companies that are interested in maintaining and improving production will invariably pay attention to training. The benefits of investment include the following:

    - The general morale of the staff is improved.
    - An improved volume of work can be expected from a trained workforce.
    - A company with a reputation for the training and development of its staff attracts better people.
    - The staff, and therefore the management, experience less frustration during day-to-day production.
    - A decrease in labour turnover can be expected.
    - Changes in technology are more easily accommodated.
    - Job satisfaction is increased.

STAFF FLEXIBILITY

Modern printing-industry conditions require an increasingly flexible workforce to operate efficiently. Human-resources skills requirements are constantly changing, and will continue to do so for the foreseeable future.

Investment in technology and high operating costs result in printing companies no longer being able to support a number of highly specialized but skill-restricted staff. Flexibility within broad production areas is required, and is becoming increasingly common. Flexibility is required to cater for:

- sickness,
- holidays,
- the introduction of new technology,
- imbalances of work within production departments,
- the resolution of redundancy situations,
- potential for promotion,
- changes in types of work.

Improved staff flexibility is achieved through carefully considered and constructed training courses. Training can take place within the company or externally, according to the type and level of training required.

SUMMARY

- Progress within a printing company can only be achieved through people. People in the form of staff, are the lifeblood of any company.
- Rapid technological change has brought demands for new staff skills.
- Experience is necessary, but must be relevant to the production situation.
- Management can now talk to, and reach agreements with, individual employees, as opposed to collective groups.
- With modern technology, more can be achieved with fewer staff.
- The key issues in human-resources planning and management are:
    - employment policies,
    - skills requirements,
    - recruitment and selection,
    - training and development,

- motivation and performance evaluation.

- The preparation of job descriptions and specifications is important.
- Interviewing for employment is a skilled operation; wrong decisions can prove to be very expensive.
- Management and production worker performance evaluation can bring considerable benefits.
- Motivation may take a number of forms. The form differs according to the employee, his/her status, and the work that he/she performs.
- Training and constant updating is very important.
- Management self-development is essential.

ACTIVITIES

- *Activity 4.1*: Describe the benefits that may be expected from employee performance evaluation.
- *Activity 4.2*: Prepare a job description and specification for a production situation in a medium-sized printing company.
- *Activity 4.3*: In what ways can production staff be motivated to achieve optimum production and quality?
- *Activity 4.4*: Describe an issue related to human-resources management that is not discussed in this chapter.

5
Printing technology

Printing technology comprises the machinery, equipment and methods used to create a printed product, and it is therefore a crucial element in the company's organizational strategy. Every printer must, at some time, consider the purchase of new technology or equipment, to replace outdated equipment and obsolescent methods, because of wear, or, most importantly, because of a change in the company's production aims and objectives. A quote that is often used in the printing industry has some relevance and truth: 'one day a printer will wake up and find the work he used to print being produced by another printer, on the equipment he could not afford to purchase', for no company can resist new production methods and the requisite new technology indefinitely.

The advance of technology and the consequent changes in the printing industry are rapid, and the result is that equipment becomes obsolete in shorter periods of time than was previously expected. This trend is likely to continue for the foreseeable future. The technology currently used in the industry is expensive, and invested capital is tied up for a long time. The installation of an inadequate or incorrectly investigated item of technology becomes a constant problem for the purchaser every time its limitations are experienced.

For companies wishing to remain in the technological forefront of production, a continuing rise in the cost of capital equipment is a major concern. Rapidly changing technology can make it difficult to assess new potential capabilities in relation to existing capacity. However, prior to any investment in technology, a number of criteria should be considered and thoroughly examined:

- an analysis of the work to be produced,
- the work category, e.g. basic, high-quality,
- the range e.g. the operation's required inline variation of operations,
- the turnaround time, e.g. the scheduled time allowed by normal production conditions,
- the work volume, e.g. the average production-run length and extent.

Consultation to obtain the opinions of those immediately responsible for production, e.g. production managers and overseers, should be considered. Many moves towards investment in new technology arise not from the production floor, but from senior management. Obviously, this is correct, and it does not indicate that senior management should surrender the right of decision. However, a production manager or overseer will certainly be aware of the day-to-day production criteria and associated problems affecting any particular section or specialist operation.

FINANCING TECHNOLOGY ACQUISITION

Capital required

If a printing company is to remain at the forefront of production in relation to quality, price and delivery, then the replacement of outdated equipment is essential. This means that finance must be ploughed back into the company, irrespective of its size or specialization. Unfortunately, reduced profit margins reduce the potential for investment in new technology using profits. Profit margins may be reduced because of difficult trading conditions, intense competition, poor costing and estimating methods, or the decreasing turnaround time required by all clients, often resulting in slack production periods, although this is sometimes offset by overtime working or premium hours. The best source for capital investment is profit. However, it is doubtful that profits alone will be sufficient to provide the level of investment likely to be required for the great majority of major items of technology. Therefore, other methods of financing must be considered.

Depreciation

Printing organizations should recognize the importance of correctly apportioned depreciation allowances as a source of finance. Depreciation on existing technology allows for replacement after a predetermined period: the period may vary from two years for computer equipment to five or seven years for large items of technology such as printing and finishing equipment. Very large items, such as web printing presses, may be depreciated over a much longer period. Another important factor is that the depreciation rate is often unable to counteract inflation, and therefore other courses must be considered.

Hire purchase

Hire purchase is often used in cases in which capital is extremely scarce. A large amount of technology has been purchased in this manner by printing companies. However, the cost is high, and it results in overall payments whose total is far in excess of the normal purchase price. Usually, about 25–30% of the purchase price is required as an initial payment, with the balance being paid over an agreed number of years in monthly or quarterly instalments. It may be that, on completion of the purchase, the technology is outdated, and has become a competitive liability.

Leasing

Some manufacturers of technology for the printing industries operate, in conjunction with finance houses or banks, a leasing scheme. The scheme allows for installation of technology in return for a monthly premium. An obvious advantage is the opportunity for a company to maintain its position in relation to technology without making a large initial investment. Leasing has an advantage over hire purchase in that hire purchase, at the end of repayment period, probably leaves the printing company with old or even obsolescent technology. With leasing, ownership does not pass to the company without additional agreement, and thus updating can be less complicated, take place at more frequent intervals, and assist a company to remain competitive. Leasing can, however, be very expensive.

Amalgamation with another printer

This, at first, appears to be a rather dramatic way of dealing with the problem of technology acquisition. However, amalgamation allows better use of existing technology, and economies can be achieved by combining a number of functions. Surplus equipment can be sold, and the finance put back into the business. An arrangement that falls short of amalgamation with another company is one in which a group of printers, each specializing in a specific production area, e.g. origination, printing and finishing, each acquire the technology that is suitable for their specialization, and then intertrade with each other. This type of operation is becoming increasingly popular, and it contributes to the substantial growth of small companies engaged in various printing processes throughout the UK.

Loans from shareholders or other interested parties

Additional finance can be raised from shareholders, clients, finance houses and banks. Good security has to be offered, and a demand for a larger share in the business is almost certain. Loans from finance houses or banks form the first charge on the assets of a business. A complete statement of expenses, probable turnover etc. is required, and the company's business plan for its short- and long-term business objectives. Also required are the management structure and strategy that have been designed to achieve the business objectives.

Mortgage of freehold property

If the company owns the freehold of property, this is a very good way of raising finance through banks, building societies or insurance companies. Selling freehold premises and then leasing them back is a further way of raising necessary finance. In this case, it is essential that the land or premises are free of any charge, and have the potential for renting for at least 50 years.

It may be possible to arrange the purchasing back of premises and land at predetermined dates.

Nonmoney-market acquisition

When dealing with the issue of raising finance, attention should be paid to the following:

- prompt debt collection,
- free money locked up in little used or unused technology,
- the reduction of the value of work in progress,
- the reduction of stocks of materials held,
- an increase in production work shifts where possible,
- the reduction of waste in production,
- the elimination of waste in administration.

Cost comparisons

Cost comparisons fall into two main categories: first, the comparison of the purchase, installation and commissioning costs of similar items of technology, and, second, the decision to replace any technology that has become obsolescent before the end of its expected working life.

To compare and evaluate the purchase costs of various items of technology, it is necessary to have an in-depth knowledge of the production capabilities of the various items of equipment. Any feasibility study relating to the installation of new technology should consider the implications for other areas of production. Ideally, the production process should proceed at a uniform pace. High production in one area can cause a bottleneck in another, e.g. the introduction of advanced printing technology may create, because of increased production, an overload in the finishing department. Alternatively, the installation of a stitching and trimming line may increase the finishing capacity beyond that of the printing department. The evaluation of alternative production methods is essential when the need to update or completely replace technology is assessed. This is, perhaps, one of the most difficult issues in the purchase of technology, as vast differences in price for finishing, printing and origination technology are evident. Price differences usually reflect the ability of technology to perform effectively to the standard required. The evaluation of work type and quality is essential when the purchase price is taken into account, as it may well be possible to purchase technology that is apparently adequate, only for it to be later found that the equipment is certainly adequate, but only for the work for which it was designed, and not for the type of work being attempted.

Caution must be exercised when one assesses the sometimes rather enthusiastic claims made by equipment manufacturers and suppliers regarding the abilities of their equipment. Some manufacturers have an optimistic view of production norms, stating the maximum speed in the catalogue, and not the average production output that can be expected when various types of material are used under changing conditions. It may be possible to produce a certain type of work on an item of technology, but the question of whether it can be produced to the standard required, to the right timescale, cost-effectively, has to be asked. Frequently, it is necessary, but difficult, to assess an item of technology in terms of its obsolescence before the end of its useful working life. Often, old technology produces to the standard required, but with production times that are long and with excessive costs. In these circumstances, the cost of existing technology must be compared with the costs of investment in new or updated technology. Hourly cost rates, manning levels and salary rates are known for existing technology, but those for proposed new technology will need to be established, and then compared with the existing rates. In certain circumstances, there is little option but to invest in new technology. Advances in wordprocessing and design technology, for example, require a company to examine carefully where it stands, or work may be lost. The same applies to a printer who owns old printing presses that are, for example, in good condition but lack electronic colour control; achieving quality within the tolerances acceptable to the client is difficult, if not impossible.

AFTER-SALES SERVICE

After-sales service comprises follow up after installation, and assistance with technical production problems that develop through lack of knowledge or incorrect working procedures. Provision for staff training also requires investigation.

The acquisition of new technology alone is insufficient to ensure improved productivity, quality and cost. Subsequent to installation, a period of induction is required, even if the company already has similar items of technology *in situ*. Follow-up liaison between the printing company and the technology manufacturer or supplier is essential. The liaison process should include the monitoring of technological performance in relation to the printer's needs, the evaluation of any operating problems experienced, and the provision of expert technical advice to resolve such problems. The initial demonstration period immediately after installation should not only prove performance ability, but also identify any changes in working procedures that are necessary, and indicate any known production practices or operations that are no longer required because of improved operating efficiency. In this initial period, operator apprehension of new technology and learning new systems, may have to be dealt with.

Even with so-called simple modern technology, the simplicity is relative. Here, the need for well defined training is important. A well designed training programme builds on existing operator knowledge and skills. It is important that training should proceed beyond the initial induction stage to the acquisition of in-depth understanding of the various operations and control processes of the technology. Invariably, an excellent return on the time and finance invested is experienced.

TECHNOLOGICAL FLEXIBILITY

Technological flexibility relates to the range of work that can be completed inline with the minimum amount of mechanical adjustment. Examples of technology flexibility are the abilities to number and perforate inline, surface or spot coat, convert printing from four colours on one side to two on either side and perfect, and use a varying number of stations on a finishing and trimming line. Increasingly, more production operations are being brought inline, resulting in a demand on technology manufacturers and suppliers to develop technology with ever increasing flexibility. The amount of mechanical adjustment that is required to accommodate varying production specifications is very important; some older technology can be adjusted to varying demands, but only in extended times. This obviously reduces company profits and/or quality, and adds to materials wastage. Production requirements and the degree of flexibility that these requirements demand from the technology should be carefully considered and analysed before purchase.

INFORMATION SOURCES

Some manufacturers supplying to the printing industries have a pedigree that is beyond question, their reputation having been built and enhanced over many years of manufacturing experience. Even so, a bewildering array of technology is offered by these and other manufacturers, and reference to catalogues and technical specifications alone is insufficient for the safe determination of a purchase. Views on the suitability of various items of technology for differing classes of work can often be obtained through visits to other printers. Also, professional organizations such as Pira International, the British Printing Industries Federation and the UK Institute of Printing may be valuable sources of information. Visits to industrial exhibitions can be useful, but caution is required, as technology is invariably exhibited and demonstrated under ideal conditions. These conditions may be difficult, if not impossible, to reproduce on the company's site. Often, as with many other things, the price one has to pay for the technology is a good guide to its potential performance.

MAINTENANCE

The expected costs of preventative and corrective maintenance must be analysed. Obviously, it is vital that all technology is maintained in optimum operating condition, and that downtime due to breakdown, waiting for parts, and priority repair must be kept to an absolute minimum. Thus the amount of the likely costs of preventative and routine maintenance for the item is very important. Obviously, these costs will vary in proportion to the age of the technology and the conditions of its use. Changing trading conditions have, to a large extent, diminished the tendency for companies to install technology, work it hard with little attention to maintenance, and replace it as soon as it becomes troublesome.

A planned strategy of preventative maintenance is recommended. Time spent in the various maintenance duties is compensated for by improved productivity and greater all-round efficiency. Also, technology that has been correctly maintained has a greater resale value. In fact technology is often judged as much on its condition as its age.

Maintenance work is of four types:

- preventative maintenance,
- breakdown maintenance,
- priority repair,
- emergency repair.

Preventative maintenance is the first essential, as it prevents interruptions to production, and thus avoids emergencies in which quality or delivery requirements are not met. Preventative-maintenance duties include the regular inspection of all mechanical and electronic units, lubrication, minor adjustments, and minor repairs and replacements where necessary. The objective is to guard against any conditions that may lead to breakdown. Preventative maintenance helps to reduce repair costs, and prevents technology becoming inoperable at the most difficult times, such as when deliveries are crucial. While a client may be sympathetic to a printer's production problems, he/she will be far more concerned about the progress of the job. Attention to technology in the form of corrective maintenance, when minor adjustments are sufficient, saves time and money. Breakdown maintenance is necessary when technology is out of commission, and it is usually necessary because of insufficient maintenance. Apart from inoperative technology and the resulting loss of finance through downtime, there are other issues: loss of opportunity, the fact that 'farming' work (the subcontracting of work to other printers) may be inevitable, and the consequent disruption to subsequent production operations. These conditions almost certainly lead to the necessity of priority repairs, requiring a visit from specialist engineers or electricians. Most manufacturers offer good maintenance provisions via planned maintenance, emergency visits, and often other planned maintenance with purchases of technology. The costs of visits at fixed or agreed times can be high, and they need to be viewed in the same way as any insurance provision. Only the company can decide on the benefits of entering into such a scheme. Historical data about maintenance and repair costs will provide a useful guide to the reaching of a decision. Remember that each item of technology is a cost centre, and therefore a profit centre.

When the overall costs of maintenance are assessed, the ability of technology manufacturers and suppliers to provide any parts at short notice should be evaluated. Downtime due to waiting for minor parts is very disruptive, and it can result in rapidly increasing costs that are out of all proportion to the item required. Routine maintenance of all technology should be the responsibility of all managers, supervisors and production operators.

ESTIMATED HOURLY COST RATES

Any successful printing company must systematically recover all the costs incurred in running the company. Any production operation is a cost centre, and, as such, it has an hourly cost rate. All the costs that can be related to an individual item of technology or cost centre must be allocated to that centre. Other costs are then proportionally allocated to the centre. When all the costs have been identified, they are allocated to the cost centres, (in this case, an individual item of technology), and divided by the total production hours available. It is almost impossible to achieve the full production potential. In fact, an efficiency ratio of approximately 80% of the possible maximum is considered to be good in the UK.

When the purchase of technology is considered, it is wise not to rely only on historical data about the costs of operating equipment, as the influence of inflation, the trading climate and business progress may mean that the true costs are not always reflected.

The ultimate objective of installing new technology is not only to provide easier production conditions, but also to ensure increased productivity, while maintaining quality at an hourly operating rate that recovers all apportioned costs, and yet allows clients to be charged competitive rates. The depreciation term and rate allocated to any item of technology plays an important role in the determination of the hourly rate to be charged.

OPERATING SPEEDS (PRODUCTION NORMS)

The maximum mechanical operating speeds given in a manufacturer's technical specifications and literature may bear little resemblance to actual production output that can reasonably be expected. Production output is influenced by:

- the work category, e.g. the technical difficulty relating to individual job requirements,
- the materials used, in terms of, for example, their quality, and their fitness for their purpose,
- the equipment's mechanical condition, e.g. its age, the conditions of use, and its design,
- the environment, e.g. the lighting and the plant layout,
- the skills and abilities of the operators, e.g. their qualifications and relevant skills,
- the temperature, heat, cold and humidity.

All these variables, contribute, at various times, to the establishment of the production norms that can be expected in any specific situation. However, great care must be taken to avoid traditional or emotive views of the possible production capacity. The fact that production norms exist does not necessarily mean that they should be accepted as correct. Technology should not be purchased and installed only to provide an easier existence for production workers; increased productivity that covers the apportioned costs and the investment must be achieved. However, increased productivity alone is not sufficient; any increase in productivity must be accompanied by an increase in profits. A company can achieve high productivity but have an unacceptable spoilage level, thus producing more work overall, but less saleable work.

MANNING AND TRAINING REQUIREMENTS

The emphasis on training for new technology should be directed towards the acquisition of manual skills; however, the development of mental faculties is also important. Correct training for new technology will result in increased operating efficiency, and improved productivity and profitability. The following points should be considered:

- Does the manufacturer or supplier provide preinstallation training, or training on-site subsequent to installation?
- Is any further or follow-up training provided?
- Is the training provided specific to the type of work being produced?
- What is the training period that is normally provided?
- Have the existing levels of experience and ability of the potential production operators been assessed and evaluated?

The trend in modern-technology design and capability is towards automation and reductions in manning levels and make-ready times. This is most obvious in the mechanization of previously labour intensive operations such as materials handling, but is also evidenced by the modular type, centralized electronic control and monitoring, and diagnosis systems currently being incorporated in the latest technology. It may be that the installation of new technology results in reduced manpower requirements. This should not be regarded as a problem caused by the purchase of technology, but as an opportunity to retrain or redeploy staff to achieve a new competitive position within the company's production field.

POTENTIAL MARKET

All companies need to develop a strategy for improving administration and productive performance. A policy of purchasing and installing technology without an analysis of the company's existing and future potential market position is inadequate, and it is unlikely to point out a clear-cut direction for company development.

Any strategy for technology purchase or updating must focus on having the most suitable equipment available for the production processes to be undertaken. Prior to any purchase, an evaluation of the technology on site should be carried out, and its relevance to changing production requirements analysed. In this way, the replacement of technology by a mere updating of the model on site will be avoided. Technology is often purchased on an emotive basis rather than as a result of in-depth analysis of the company's requirements in the short-term and the long-term.

The purchase of any technology should bring increased productivity. However, a major problem often exists in the printing industry, where because of very tight delivery dates, and consequent changes in work-programming priorities to achieve customers' demands, the technology may experience periods of work overload, and at other times have very little to do. Therefore, the market should be carefully examined. Specialization in one type of work or production operation is becoming increasingly popular. Such specialization allows very competitive prices, resulting in increased workflow. However, there is also a disadvantage: a company choosing specialization is even more subject to changes in technology and methods than one that offers a wide range of production operations. Examples of such specialized operations are repro houses, trade printers and trade finishers. Prior to any investment being made, some questions should be answered:

- What is the demand, present and future, in terms of type and volume, of work for the existing technology?
- What is the potential of the proposed new technology?
- Can an increase in work be realized through existing clients? In other words, is the maximum possible amount of work being obtained from each of the current accounts?
- Does the proposed new technology allow movement into a new field of production? If so, what is the potential for attracting new work?
- Can the existing sales force introduce sufficient work to meet the capacity created by increased productivity?

Any situation or circumstance that means that there are negative answers to any of these questions should be resolved, as a printing company can only stay in the forefront of production by installing modern technology and adopting up-to-date production methods.

DURABILITY

An analysis of durability as it is related to the expected economic life of technology is essential. Rapid technological advances may mean that equipment has years of productive life left, but has become economically disadvantaged. Old, but well maintained, technology can very often produce excellent quality, especially in the hands of experienced operators. However, the time taken to achieve the required standard is nearly always prohibitive, and it leads to either lower profit margins or higher charges to the clients. These situations will not allow a company to progress.

In this context, durability means the resistance to the everyday wear and tear that occurs as a result of production operations. The strength of the construction of the main frame and the operating units should be consistent with the expected usage. The reliability of the electronic components is also very

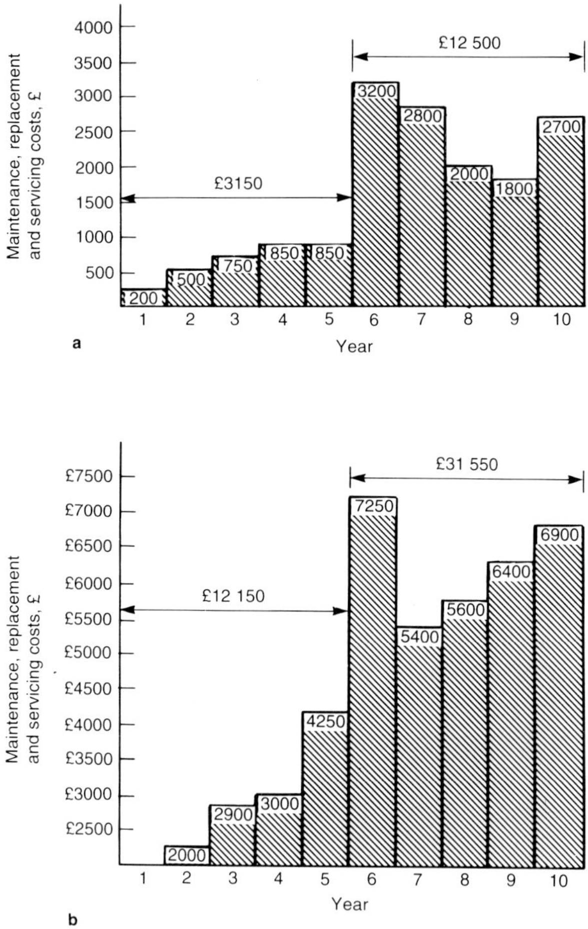

Figure 5.1 Maintenance, replacement and servicing costs for two items of technology. [The availability of spaces and obsolescence should also be considered; (a) purchase and installation costs: £57 000, (b) purchase and installation costs: £182 000.]

important. Technology of light construction is generally not as capable of resisting the rigours of production as that of more robust construction. This may appear to be obvious. However, it is possible to find wide variations in construction design and strength in items of technology designed to produce identical work. The durability of any individual item of technology is influenced by its method of use. However, even with careful and correct use, the active life varies in relation to the amount of maintenance, routine and emergency, that equipment is likely to require and receive. The rises in operating costs related to durability in productive use and, the maintenance costs for a 10-year period for two items of technology used on double day shifts are shown in Figure 5.1.

It can be seen that, normally, the maintenance costs incurred during the period following installation are relatively low, and then show an acceptable increase over the next 6–10 years. However, a sharp rise in operating costs can be detected for replacement parts and servicing over the next 6–10 years. This is mainly due to wear of critical parts or units. Overhauling or rebuilding the technology will create a decline in the operating costs. However, it is unlikely that the operating costs, the maintenance and repair costs, will ever return to the level of the earlier years of the equipment's life.

The durability of computer technology, in view of the rapid advances in this area, is probably less significant than that of other large items such as multicolour presses, both sheet and web, and stitching and trimming lines.

NOISE

Noise related to the production operations of single or collective items of technology is very important, and influences productivity and possibly operator health. Technology manufacturers and suppliers appreciate the effects of noise on the individual and the environment, and constantly carry out research to reduce the noise levels produced in operation. Noise in a printing company may be defined as unwanted and unnecessary sound.

Even in large production areas noise levels can be contained below the nuisance level by correct guarding and shielding, for example by installing hoods over folding equipment, isolating compressors, and sound proofing areas of intense mechanical operation and therefore noise production. Constant exposure to noise, however, affects hearing, and can lead to varying degrees of deafness. Noise-induced deafness can be temporary, lasting for 14–16 hr, or permanent.

NOISE AND VIBRATION

The effects of noise on cognitive performance are also constantly being researched, and there is controversy about the effects that various levels of noise can have on performance. However, noise clearly has an effect on overall performance, as increasing levels of noise require an increase in operator concentration for production operations to be carried out efficiently. Lowering of resistance to frustration, reductions in comfort and increased fatigue may also be evident. However, the degree of these effects will vary from individual to individual. Closely associated with noise is vibration. As with noise, the effects of vibration vary from person to person, but it is not considered to affect cognitive performance. Modern technological design has, to a large extent, eliminated vibration, but older equipment that is inadequately maintained and installed on inferior bases or support not only produce vibration, but can also contribute to building-fabric damage.

The effects of background noise, such as radios, are disputed. Loud music is often wanted noise, and, if the noise is enjoyed by the worker, then it is doubtful that any bad effects on the individual will be seen. However, to some individuals, loud music can be an annoyance. An important consideration is, therefore, the degree of annoyance to the production and administrative staff overall. The wearing of individual radio headsets should be discouraged, as it hinders effective verbal communication, and the potential for accidents increases.

The effects of noise are not related to the internal environment of the printing company alone. Neighbouring industries and residential areas must be thought about, and they will of course be subject to health and safety and environment law. The installation of technology that creates excessive noise or vibration may mean that a restriction in the hours during which it can be operated will be imposed by the local authority.

DESIGN

The design of technology has implications for the operator's workspace as well as for the space occupied by the production area. The extent of the movement necessary to achieve essential functions, and the ease of access to mechanical units, services etc. are affected by the design. Machine operators may have achieved a high level of skill, but performance levels decline when operators are placed under physical stress because of constant and often restricted movement. Attention to technology design in relation to appropriately positioned controls, information displays, movement required, and the ease of access to various operating units, is essential.

The design of technology affects two main areas; first, the ability to create the desired product through high-performance equipment, and, second, the demands placed on the operator's movements that are required to achieve production. The principles of the suitability of the design in terms of the creation of the end product are mainly determined by the type (category) and standard of quality of the work to be produced. Obviously, modern technology, which benefits from research and development and stringent manufacturing requirements, is far more likely to achieve the standards required. The purchase price of technology may be a guide to the efficiency that can be expected.

Technology designed for a specific purpose can be purchased over a wide price range. However, that which is lower-priced may be inferior in some aspects, and consequently offer less opportunity for effective production. This means that there is less provision for quality achievement, in terms of both mechanical design and operator movement. Careful analysis of the work specification is essential to ensure the suitability for the type of work expected. Evaluation of the work produced by the technology can be a valuable guide.

The effects of the technological design on the operator is a critical issue when new or previously used machinery of any description is purchased and installed. An important factor is the grouping of the various controls that it is necessary to use to operate the equipment effectively, as is the movement required between the various control units to produce effectively under normal operating conditions. This is not an easy problem to solve, as consideration of the design appearance alone is insufficient. Operator comfort and safety must be prime considerations. Comfort can be defined as easy access, normal body movements in operation, and the grouping of necessary controls to avoid unnecessary travel during preparation, setting up and production. Safety can be defined as the adequate guarding of all moving parts without the creation of operating difficulty through the restriction of access to essential areas. Other considerations are as follows:

- *Frequency of usage of controls*: The more frequently a control is used, the nearer it should be placed to the prime operation position. Large technology that covers extended areas should have duplicate controls on all units and stations. These controls should allow the operator to override other controls or isolate them, thus ensuring operator safety.
- *Function and time sequence of operation*: The operation of the majority of the technology used in print production usually takes place in a predetermined sequence. If the controls are arranged with this sequence in mind, it will reduce unnecessary movement and the possibility of mistakes.

Each advance in technology results in ever more complex control systems that take over functions previously carried out manually. Research and development aimed at reducing operator time and movement, especially for large technology, is focusing on fully digitized control systems that replace electromechanical systems. Fully digitized systems provide, at one convenient operator position, information relating to:

- continuous functional diagnosis of the overall operation, during which incorrect settings or functions are immediately obvious without the need for constant movement and inspection,
- precise areas of malfunction, thus allowing immediate correction, with the benefit of waste reduction.

ANCILLARY EQUIPMENT

Ancillary equipment consists of integrated or optional extras associated with items of technology. Examples are printing presses with facilities for numbering and perforating, inline coating, folding machines with slitting or gluing facilities, stitching and trimming lines with cover insertion, and web presses with sheeting or rereeling. The key to decisions relating to any necessary extra inline operations is a very careful evaluation of the work type and the consequent manufacturer requirements. Errors of judgment about demands on production technology, caused by misunderstanding or lack of research and evaluation, may be very expensive every time a particular production situation arises in which additional operations require separate production runs. Therefore, do the following:

- Evaluate the work type and category.
- Determine any process or operation that may be regarded as ancillary.
- Determine the frequency of ancillary operations.

- Establish the percentage of the overall turnover contributed by the work.

An analysis of the answers to these points will provide a clear indication of the need either to adapt existing technology or install new technology.

FLOOR SPACE, REQUIRED POWER, AND OTHER TECHNICAL SPECIFICATIONS

The floor space required by the technology has implications for:

- the rateable value of the floor space,
- manning levels,
- operator movements,
- the areas required for work in progress and storage between operations,
- the designated working space and benches, chairs, tables etc. required to achieve efficient production.

Accurate estimation of the floor space required is perhaps the most basic consideration in ensuring that the technology will be able to produce at optimum levels. Adequate passageways and means of access and egress are a matter of legislation. However, working to the minimum imposed by the law is unproductive. Distances between items of machinery need to be sufficient to allow freedom of movement without contortion. Adequate freedom of movement all round an item of equipment or machinery is essential. The practice of placing equipment close to walls, although it reduces the demand for floor space, almost certainly results in restricted movement on one side of the equipment. Operator annoyance, frustration and downtime can result. Thus, floor-space calculations should allow for adequate movement round the whole area occupied by the equipment.

Another critical issue is the strength of the floor. Before any installation, the weight and distribution of the equipment, and the weight of work in progress and stored materials, should be carefully assessed, preferably by a qualified structural engineer. Insufficient strength of the base or the foundations will lead to long-term problems such as impaired mechanical efficiency and increased vibration, and there will need to be frequent corrective maintenance by engineers. Loss of production quality is likely, and, in extreme cases, production will have to cease pending the resolution of the problem.

The evaluation of technology should also include any specific temperature, air conditioning or illumination requirements. Many printing operations are carried out under specific temperature or humidity levels, and they may require precise illumination before optimum production can be achieved.

Power loads should be assessed in relation to existing provision. Particular attention should be given to provision for computerized and electronic control systems.

Ease of access for technology installation is very important, and alternatives may have to be assessed. The dismantling of assembled mechanical units, or structural alterations to buildings, can be very expensive. Ideal installation conditions will allow adequate access to a ground-floor site for erected mechanical units. Installation on floors above ground level can add considerably to costs.

NEW OR USED TECHNOLOGY

Decisions about the purchase of technology nearly always involve the question of whether it should be new or used (second-hand). Some companies have a policy of only installing new equipment. This policy is perhaps the wisest where it is possible. However, many printing companies are small, and have restricted financial resources. For these companies, the high financial investment required for the installation of new technology is prohibitive, and used equipment is the only route available.

The view that all used equipment is inferior is false. Well maintained used equipment can be, and often is, in better condition than recently acquired equipment that has not received correct care and maintenance.

Important points that must be taken into account when the purchase of used equipment is considered, in addition to those relating to the purchase of new equipment, are:

- its age (this must be checked and verified by the serial number),
- its method of use (single, double or treble shifts),
- historical data relating to accident or repair, and any maintenance records (refer to the life clock if required),
- its previous owners and the type of work produced,
- the extent to which the equipment will have to be dismantled and rebuilt on site,
- the ease of egress from the existing site, and the access to the new site,
- the availability of any parts required,
- whether it can be seen in production,
- whether it is available for testing before purchase,
- the reputation and experience of the seller, if it is to be purchased through a dealer.

The used-technology market in the UK is extensive, and the market is increasingly encompassing large areas of continental Europe. International trade in used technology is common, and enquiries to dealers and agents may find the requirement being met in Germany, France, Denmark, Sweden etc., rather than in the UK.

There are therefore additional points to be considered:

- shipping and insurance,
- the cost and time involved in viewing the equipment, and the question of who pays for this,

- variations in exchange rates.

The question of training staff when this is required is more problematic when used technology is purchased. Many dealers will install the equipment, demonstrate it, and consider the contract fulfilled. Additional training may be possible, but at extra cost. Also it may be that the seller is unable to offer training or other assistance. This will certainly be the case if technology is purchased directly from another printer.

Finally, it is necessary to consider the record and reputation of the dealer or supplier. Simple cleaning and repainting of the equipment for resale is insufficient. Reputable suppliers will dismantle, overhaul and rebuild where necessary. Investigation into a supplier's methods of operation is advised, and a visit to the work site should be made if appropriate.

PURCHASE AND INSTALLATION FACTORS

It can be seen that the purchase and installation of technology is a very important and complex issue that requires careful consideration of many factors including:

- the capital required,
- cost comparisons,
- maintenance,
- after-sales services,
- flexibility,
- operating speeds,
- the satisfaction of current users,
- manning and training,
- estimated hourly cost rates,
- potential markets,
- durability
- noise, design and ancillary equipment,
- floor space and power,
- depreciation and resale values.

RECORDS

The preparation of a comprehensive record of all the technology within a printing company is recommended. Benefits derived from the keeping of efficient records include the following:

- Accurate data about the life of the technology is immediately available.
- A basis is provided for the development of maintenance plans.
- Factual information about the true operating costs is provided.
- Areas or items of technology that require abnormal maintenance are identified.
- Valuable time is saved when technology serial numbers or part numbers are required.
- It assists with depreciation calculations and decisions on obsolescence and writing off.
- Accurate information is provided when the causes of problems are assessed.

To be of value, all information has to be accurate and reliable, and recorded as early as possible. Reliance on memory, in the absence of records, can be misleading: concepts of time, events and actions taken can alter considerably with the passage of time. An example of a technology register is shown in Table 5.1.

SUMMARY

- The selection and purchase of technology is a key issue in any printing company, irrespective of its size.
- No company can resist the introduction of new technology indefinitely and still remain competitive.
- Rapid advances in printing technology result in obsolescence sooner than ever before.

Table 5.1 Technology register

Description of item	Serial number	Number of items	Year purchased	Estimated life (from purchase), years	Current replacement cost, £	Maintenance action	Cost, £	Date
Heidelberg MOZP	123546	1	1989	7	158 000	Adjustment to perfecting mechanization	380	11/90
						Retime feeder	235	3/91
MBO folding machine	654213	1	1985	10	48 000	Adjustments to feeder	175	4/92
						Add hoods	700	3/89
						Change rollers	1250	7/90
						Overhaul	2600	5/91
Computerized cost and estimating system	Apple Mac	2	1990	4	21 000			
Perfecta guillotine	Seypa 115		1987(1)	10	42 000	Adjustments to hoists	394	1989
						New blades	400	1993

A separate sheet is used for each item of technology. Records can be maintained manually or on computer.

- Printing-industry technology is very expensive, and capital invested in it is tied up for long periods.
- Any mistake in selection will become obvious in certain production situations.
- A careful and precise analysis of the category of work to be produced, and the mechanical and electronic demands to be made on technology, are essential.
- Consultation with the production managers and/or overseers is recommended.
- A correct evaluation of the market potential, manning levels and expected productivity should be made prior to the decision to purchase.
- Various methods of raising capital to invest in technology are available. The correct method will vary, according to the company's circumstances.
- Finance may be raised by improving administrative and production methods within the company, and thus reducing demands for external finance.
- Reliance should not just be placed on manufacturers' and suppliers' claims about performance; other sources should also be used.
- Any assessment of technology must take staff and training into account.
- Every installation of new technology must bring about improved productivity.
- Ergonomic considerations are important.
- Decisions as to whether to use new or previously used technology are required.
- At least 16 points must be considered before technology is purchased.
- Accurate records are essential.

ACTIVITIES

- *Activity 5.1*: Identify the implications for the staff of the installation of new technology.

- *Activity 5.2*: Identify and evaluate the advantages and disadvantages of installing new technology as against those of installing previously used technology.
- *Activity 5.3*: Explain the effects of inadequate maintenance on production and human resources.
- *Activity 5.4*: Describe an issue related to printing technology that is not discussed in this chapter.

6
Print production and operations management

Printing-company management duties are those of achieving business aims, and objectives, and performance, social and profit targets.

To facilitate this, management responsibilities are often divided into four main areas:

- financial management,
- sales and marketing management,
- human-resources management,
- production management.

Financial management

The financial management has an overall responsibility for the economic and financial position of the company, the funding of capital investment in new technology, administration expenses, wages and salaries, capital for materials, the flow of working capital, costing, budgetary control, the annual accounts, capital project planning, profits etc.

Sales and marketing management

Without effective marketing and sales management, a company will probably decline and eventually die.

Duties include control of the sales operation, pricing (in conjunction with financial management), sales promotion, estimation, advertising, developing marketing opportunities, new product development, planning and measurement performance, market research, and the determination of the best method of distribution.

Human-resources management

Human-resources management duties are the establishment of human-resources requirements and skills, staff levels, recruitment and selection, industrial-relations training, the working environment, job evaluation, and staff appraisal.

Production management

This chapter deals with the production-management function. The main production-management duty is to achieve the performance targets set by the overall management team by ensuring that the production of the right quantity and quality of printed work at the right time at minimum cost is maintained. This can only be achieved through research and market positioning. The advantages to be gained are:

- price leadership through the effective production of printed products at lower prices than those of the competitors; however, this must be achieved without any reduction in product quality,
- product specialization, in that the company achieves a reputation for quality, reliability and a helpful approach to clients.

PRODUCTION MANAGEMENT LINKS WITH OTHER AREAS

Production management cannot work in isolation from other management areas (see Figure 6.1). All production decisions

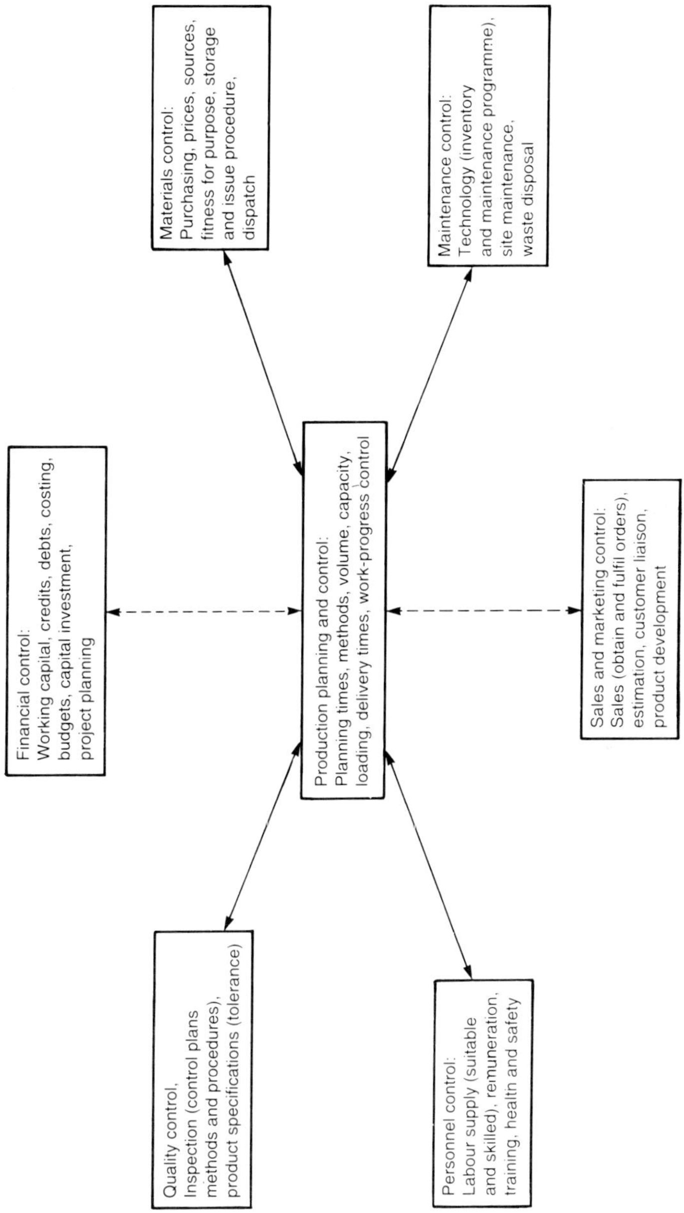

Figure 6.1 Production management: direct and indirect links with other management areas

will have implications for other areas, e.g. a change in the production technology scheduled to be used for a particular job, perhaps because of overload or the technology being out of service for a period, will almost certainly have a financial impact on the estimated prices. The same situation applies to unforeseen overtime demands when the extra costs incurred cannot justifiably be passed on to the customer. Obviously, these circumstances will have implications for sales and marketing also.

PRODUCTION PLANNING AND CONTROL

It is not possible to lay down rigid rules for the planning and control of print production. Each company will need to devise a system to suit their individual circumstances, taking into account the type of work, the technology in use, turnaround times and quality requirements.

Production planning is defined as a system that foresees each step in a series of production operations, and arranges for each operation to be achieved in the right time with the greatest possible efficiency.

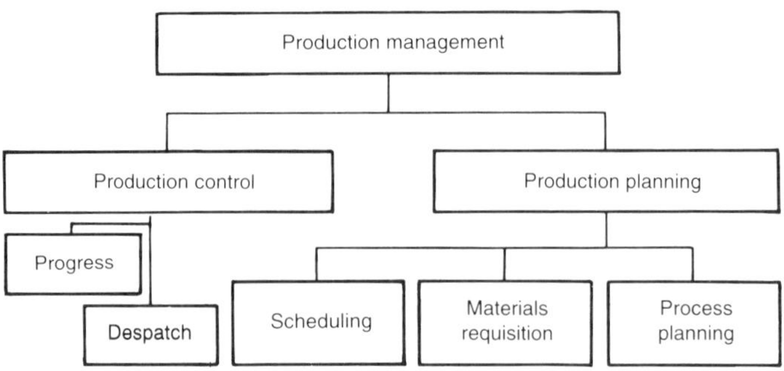

Figure 6.2 Production management

Production control is the process of action and initiation of the production stages, and constant feedback about the work situation (see Figure 6.2).

When work is correctly planned and controlled, better use of the potential capacity will be achieved, delivery times can be more easily met, and urgent work can often be accommodated without disruption to other production.

No matter what type of planning and control system is adopted, they all incorporate common principles, as follows:

- They plan production methods and processes.
- They determine the sequence of production operations.
- They provide a factual estimate of the time required and the time available to complete the work.
- They allow the collection and recording of data about the production progress of individual jobs.
- They ensure that customers' technical work specifications are achieved.

These objectives are achieved through a sequence of operations:

- *preproduction planning*: the determination of the production methods, operational times, technology and department capacities, and materials required,
- *information gathering*: the preparation of work specifications, relating to the work type, the extent, specialist operations requiring external production, delivery dates, etc.,
- *scheduling and loading*: the establishment of production times, production methods and processes, the technology to be used, and the sequence of operations,
- *implementation*: the completion and issue of production specifications to production areas,
- *progressing and controlling*: the monitoring of the progress of work in relation to the scheduled position at a given time or date, and adherence to quality standards.

The carrying out of this sequence of operations should result in efficient production control in relation to:

- the provision of all the materials on the dates on which they must be available, ensuring their fitness for their purpose,
- the setting of realistic times required to complete the work, and the time available for production under existing circumstances,
- the determination of the most suitable technology, its availability, and possible alternative technology if required, and the capacity at each work or technology centre and any vacant capacity,
- the provision of human resources at any given time, and a clear indication of any movement required,
- the effects that any amendment to the production schedule may have on other work in progress,
- the identification of any work that appears to be likely to exceed estimated figures, thus allowing the selection of alternatives at an early stage.

PRODUCTION REQUIREMENTS AND SEQUENCE

The great majority of printed work begins with the design phase, following the identification of customer requirements and a production enquiry. Work may have its origin in and be designed on the customer's premises, or on those of a design company or publisher. It is here that a difficulty may arise: orders are usually carried out in response to customer demand, and they therefore have an individual customer specification. This results in production being job-controlled, as the combination of a number of printing jobs is rarely possible.

Production schedules may also take into account customer priority and flexible delivery, because not all print production is required at the same time. However, in other cases, failure

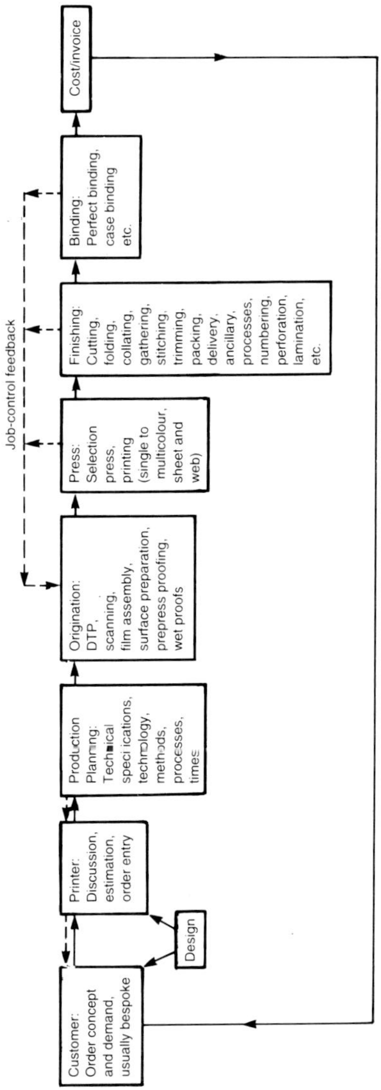

Figure 6.3 Process chart for printed work

to deliver work on time may result in penalties and certainly customer dissatisfaction.

Thus an intelligent and factual forecast of production times and costs must be made before any production begins. There must therefore be a link between production management and the sales staff.

The production sequence for a typical print order is shown in Figure 6.3.

Following the placing of an order by a customer, an assessment of each production operation and its complexity is required. However, a sound knowledge of printing and allied processes is essential before an intelligent understanding can be reached and evaluation made of each job in terms of the production plan, methods, processes and materials.

The following list gives some examples of areas in which decisions must be made in relation to the work in hand:

- *Design*:
  - concept,
  - roughs,
  - visuals,
  - artwork,
  - photography,
- *paper and board*:
  - quality,
  - weight,
  - size,
  - caliper,
  - grade direction,
  - trim allowances,
  - special makings,
  - availability and lead times,
  - numbers of copies,
  - delivery deadlines,
- *composition and DTP*:

   - class of work,

     - bookwork,
     - periodicals and magazine,
     - general and jobbing,

   - copy prep and technical planning,
   - electronic composition,
   - manual and electronic page makeup,
   - bromide, negative or positive film,

- *reproduction and printing-surface preparation*:
   - quality of originals,
   - artwork,

     - camera ready,
     - quality,

   - quality of transparencies,
   - colour separations,
   - camera/scanning operations,
   - layout and imposition,
   - planning and film assembly,
   - printing down, and numbers and sizes of plates,
   - proofs, photomechanical or press,

- *printing*:
   - process,

     - lithography,
     - screen,
     - flexography,
     - photogravure,
     - letterpress,
     - sheet or web,

   - methods,

        - half-sheet,
        - sheet-perfect,

    - extent: lengths of runs,
    - colours: colour process, or self colours (pres),
    - sequence of printing,
    - press times required,
    - ancillary operations,

        - numbering,
        - perforating,
        - coating,

    - inks to laminate or varnish,
    - specials,

        - foil block,
        - ink jet,

    - selection of technology,
    - inks,

        - light fast,
        - bleed-free,
        - odour-free,

- *Finishing and binding:*

    - cutting,
    - trimming: pamphlets and magazines,
    - stitching lines,
    - perfect binding,
    - folding,
    - cutting and creasing,
    - gathering and collating,
    - stationery or account binding,
    - sewing, interleaving, numbering,

- ○ case binding,
- ○ packing methods,
- ○ delivery methods.

Orders and instructions

Printing companies will have an area for the reception of orders. Depending on the type and complexity of the work, customer liaison may be through the works or production manager, or perhaps an individual department overseer in the case of a special technical requirement. When an order arrives in the company, it may be accompanied by:

- an order,
- technical instructions, in part or in full,
- materials in the form of copy, transparencies, prints, various stages of artwork etc.

Instructions may sometimes be given verbally. In this instance, written instructions should be obtained as soon as possible. Verbal instructions, particularly in the case of print production, may not be interpreted in the same way by different people. This may lead to misunderstanding, wasted time, extra costs and general frustration.

When accepting work from a large organization with a number of purchasing authorities, it is wise to obtain the name and address of the customer and details of to whom the work and invoice should be sent. At the same time, clarify any promises relating to proofs and dates for delivery. Any printing company that values its reputation must keep to its production promises, and thus every care must be taken to ensure that no promise is made that cannot be fulfilled. On receipt of a printing order and materials in the form of artwork, etc., the first action following initial inspection and discussion must be to copy and file all the relevant order instructions. Artwork and other materials are then allocated to the relevant production areas. Any production operations that require subcontracting,

WORKS ORDER NO: ....

CUSTOMER: .... CUS. CODE: .... RECD: ....

JOB TITLE: .... PRO. CODE: .... NO: ....

CUSTOMER CONTACT: .... TEL. NO: ....EST. NO: ....

DESCRIPTION: ....

....

QUANTITY: .... SIZE: ....

COLOUR OF INK: .... MATERIAL: ....

COVER-COLOUR OF INK: .... MATERIAL: ....

NUMBERING/PERFORATING INSTRUCTIONS: ....

BINDING INSTRUCTIONS: ....

PREPARED BY .... APPROVED BY ....

DESK TOP PUBLISHING

PROOF REQUIRED YES/NO   CORRECTIONS YES/NO   PLAN DURATION: ....

PAGE/FORM SIZE ....

PROOF DATE OUT	PROOF DATE RETURNED	DESIGN YES/NO			
1.	1.	ARTWORK/PREPARATION YES/NO			
2.	2.	COLOUR SEPATATION YES/NO			
3.	3.	DATE		DATE	
4.	4.	A/W OUT	C/SEP.OUT	A/W IN	C/SEP. IN
1. ARTWORK TO BE KEPT YES/NO FILE NO:					
2. FLOPPY TO BE KEPT YES/NO FILE NO:					

COMMENTS/REMARKS ....

FINAL PROOF APPROVED BY .... ON .... CHECKED BY ....
SIGN

HANDED OVER TO REPRODUCTION SECTION

DATE: .... SIGNED SUPERVISOR/SENIOR TECHNICIAN ....

REPRO/PASTE UP

COLOUR SEPARATION REQ. YES/NO QUANTITY ....

H/TONES .... DIRECT REPRODUCTION ....

CONTACTS NEG/POS REQ. .... PAGES/FORMS PER SHEET ....

NO. OF LEAVES/PAGES/SETS .... TO PADD/STICH/MINA/PERFECT/SADDLES/STAB

TYPE OF PLATES NEG/POS .... AMOUNT OF PLATES ....

COMMENTS REMARKS ....

HANDED OVER TO PRINTING SECTION   DATE .... SIGNATURE ....
SUPERVISOR/SENIOR TECHNICIAN

Figure 6.6 Works order for a large printing company

PAPER PREPARATION

TYPE					
COLOUR(S)					
STOCK SIZE		mm	X		mm
PRESS SIZE		mm	X		mm

MATERIAL REQ. NO. ....

AMOUNT PER SHEET ....

QTY. .... BOOKS/PADS.......... QTY. PER SETS ....

PRINTING

SPECIAL PRINTING INSTRUCTIONS ....

....

COLOUR INK .... PROCESS INK SET ....

% OVERS .... NUMBERING INSTRUCTION ....

M/C UTILIZED .... PERFORATED ....

PROOF TO BE CHECKED BY CUSTOMER BEFORE FINAL PRINTING YES/NO

PLATE TO BE KEPT YES/NO FILE NO. ....

PRESERVED UNTIL AFTER DELIVERY YES/NO

COMMENTS/REMARKS..........

HANDED OVER TO BINDING/FINISHING SECTION DATE .... SIGN. ....
SUPERVISOR/SENIOR TECHNICIAN

BINDING/FINISHING

COLLATING INSTRUCTIONS .... FOLDING INSTRUCTIONS ....

COVER .... QTY PER SIGNATURE AFTER FOLD ....

ADDITIONAL BINDING INSTRUCTION ....

....

FINISHED SIZE ....

COMMENTS REMARKS..........

HANDED OVER TO DESPATCH CLERK DATE .... SIGN. ....
SUPERVISOR/SENIOR TECHNICIAN

DELIVERY RECORDS						
QUANTITY DELIVERED			SERIAL NO. BALANCE		SIGNATURE DESPATCH CLERK	REMARKS IF ANY
DATE	DEL. SHT. NO.	AMOUNT	FROM	TO		

Figure 6.6 continued

Other systems may be devised by printing companies to suit their particular production requirements. The degree of detail entered in the instructions will depend on the work type and complexity.

An example of a basic works order is given in Figure 6.4. A typical works specification sheet is shown in Figure 6.5. A works order for a large printing company is shown in Figure 6.6.

After identification and the entry of work specifications on a works order, production planning and scheduling can commence.

Production planning and scheduling

The objective of the planning and scheduling operation is to allocate the technology and human resources available, with the aim of achieving the maximum operating efficiency that is consistent with quality and price requirements. With print production, there are two main types of demand for planning:

- *Batch working*: Batch working has the following characteristics:
    - Each job is individual, and has its own specification, and there are wide variations in production demand and sequence.
    - Part delivery may be required.
    - Completed or part-completed work may have to be stored.
    - Waste can be high, for the above reasons.
    - Machine loading can become a complex operation.
    - Very detailed work specifications are often necessary.
- *Continuous production*: This is the type of production planning and control that is used with ongoing production, as in magazine and newspaper work. Methods are usually

standardized, with few production variations and an even flow of work.

Some basic terms used in production planning and control are as follows:

- *activity*: any part of a job which occupies time and is necessary to the completion of the work,
- *event*: the beginning and end of a production operation or group of operations,
- *float*: the difference in the time available and the time required for an operation,
- *total work time*: the whole of the time required to complete the work, including waiting,
- *critical path*: the sequence of activities which determines the total job time, taking into consideration concurrent and consecutive operations.

A production schedule may comprise a simple list or sequence of operations, some concurrent and others consecutive, depending on the nature of work, or it may be a wide-ranging schedule that incorporates a large number of production details. In some cases, a critical path may have to be introduced. The production of a series of books with specific publication dates may require this type of planning.

The production situation is often displayed on a bar or graph chart, and this has the following advantages:

- Individual resource demands are clearly displayed.
- There is a physical expression of periods for which no work is programmed or available.
- The work situation is easily understood by any member of staff.

Various types of chart are used for manual operation:

- an arrow diagram,

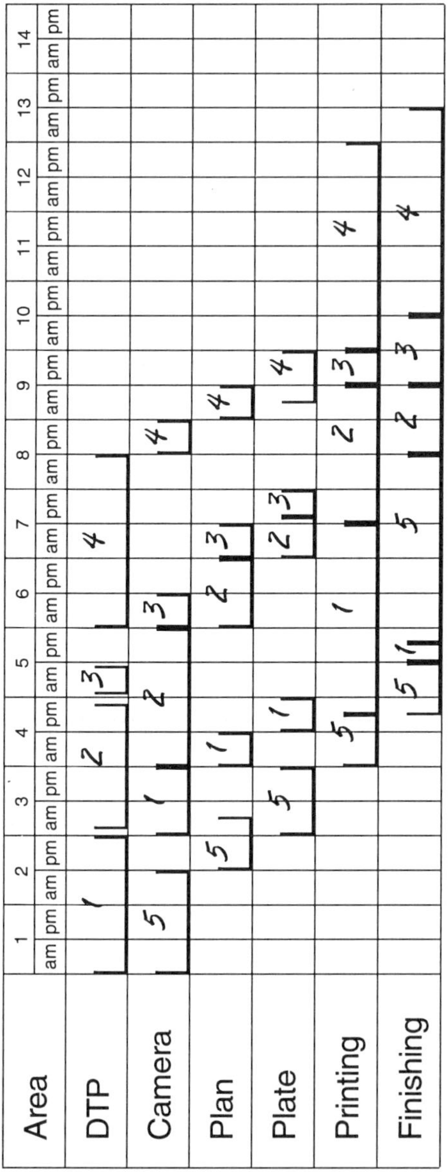

Figure 6.7 Basic chart used by small printers

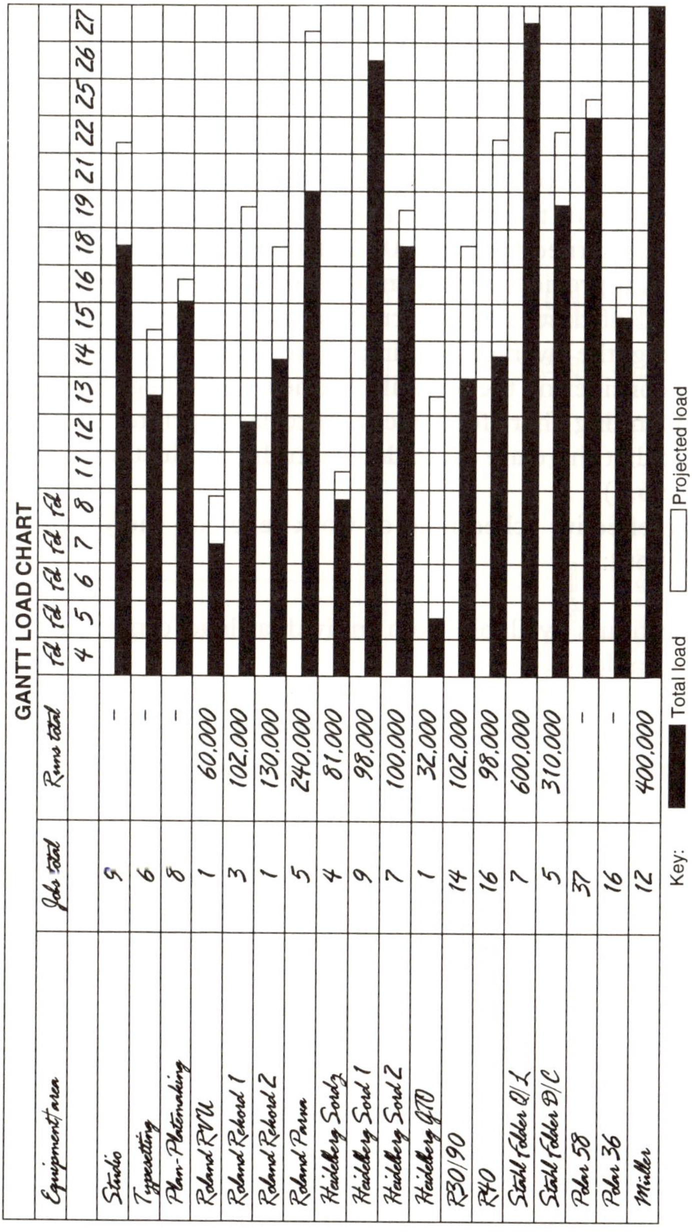

Figure 6.8 Chart showing technology loading over a period

Table 6.1 Weekly production performance analysis

Machine	Hours available	Run-on productive Hours A	Make-ready productive Hours B	Total productive Hours A&B	Maintenance	Waiting	Other duties (assist with platemaking etc.)	Productive time, %	Comments
MOVP	54.50	44.75	4.00	48.75	2.00	3.75	—	89.50	Good result
Komori	36.00	26.75	1.25	28.00	5.50	2.50	—	77.80	Query maintenance time and reason
GTO	40.00	27.00	5.50	32.50	2.00	4.00	1.50	81.30	
Stahl	35.00	26.00	3.00	29.00	—	—	—	82.28	
Total	165.50	124.50	13.75	138.25	9.50	10.25	1.50	Average 82.70	

Table 6.2 Analysis of production in relation to capacity

	Basic production hours available	Actual production hours recorded	Downtime hours	Overtime hours
Week 1 (2–7):				
Origination	284	326	40	42
Press	80	184	16	104
Finishing	94	136	72	40
Total	458	320	128	186
Week 2 (9–15):				
Origination	166	226	66	60
Press	106	232	20	124
Finishing	154	166	30	10
Total	426	624	116	194
Week 3 (16–21):				
Origination	232	290	16	58
Press	64	170	8	104
Finishing	134	168	40	34
Total	430	506	64	196
Week 4 (23–28):				
Origination	236	308	16	58
Press	132	208	2	74
Finishing	150	152	58	2
Total	518	668	76	134
Department summary:				
Origination	918	1150	138	218
Press	382	794	46	406
Finishing	532	622	200	86
Total	1832	2566	384	710

- each production worker's salaried time for the previous month,
- the overtime recorded by each production worker for the previous month,
- the downtime for each worker for the previous month,

- data about production methods in relation to production efficiency and materials usage.

Analysis of these criteria is a specialized function. However, all print managers should recognize the importance of this information when it is used as a basis for improved management control. Methods used to collect and record the information often come under the heading of work study. Work study is a generic term that is used to describe method study and also work measurement. A third area, process study, is sometimes added. Factual recording of what is being achieved on a specific job or operation by method or work study is a scientific way of analysing the total work situation in order to determine the most efficient overall work systems. Examination and study is carried out by the asking of a series of questions. Often this is a management function, but production workers can be involved, and they often have valuable information and insight into the situation being studied. The types of questions asked are shown in Table 6.3.

All the means, including human-resources and technology, by which production is achieved are examined. Thus the study includes the questions of:

Table 6.3 Examination of methods

	Present methods	Possible alternatives
Purpose	What is achieved? Is it necessary?	What else could be done? What should be done?
Place	Where is it done? Why there?	Where else could it be done? Where should it be done?
Sequence	When is it done? Why then?	When else could it be done? When should it be done?
Person	Who does it? Why that person?	Who else could do it? Who else should do it?
Means	How is it done? Why that way?	How else could it be done? How should it be done?

- human-resources utilization,
- technology and equipment utilization,
- plant layout and ergonomics,
- materials management,
- process efficiency.

Work study, when carried out correctly, should lead to improved productivity and profitability.

Procedure

The basic work-study procedure is as follows:

- Identify the job or operation to be observed and studied.
- Record all the facts about existing methods and processes.
- Analyse all the facts using an unbiased approach.
- Select the most suitable option for practical and cost-effective application.
- Implement the selected method or process, and maintain it using control and frequent checks.

The use of an activity chart in conjunction with daily work dockets, analysis sheets and work-study results will provide a valuable source of factual information on which decisions about possible alternatives to existing methods and processes can be based.

Work measurement

Work measurement is a function that is complementary to work study, and it is designed to determine the most effective ways of producing work, and to eliminate waste of effort and time. It is not intended to speed up production artificially or exploit the production worker. Correctly applied work measurement tends to reduce production time, and this is sometimes misconstrued as workers acting faster.

ACTIVITY CHART

Equipment/Machine ______ Location ______ Date ______ Number ______

Operator/s Name/s ______

Operation and work details	Staff employed on activity 1 2 3	Time Taken	Comments

Figure 6.11 Basic activity chart

For any specific job study of the time taken to achieve the various operations, a production worker of average ability, working with average effort, under normal production conditions with normal technology, should be selected. The whole aim of the study should be to suggest improvements to production processes and times. A typical activity chart is shown in Figure 6.11.

INTERRELATIONSHIP BETWEEN HUMAN RESOURCES AND TECHNOLOGICAL RESOURCES

Achieving efficient production is not a matter of planning and control alone; relationships between technology and people have very important implications for production. For the printing industry, especially in production operations, human and technological considerations are interdependent; neither can be dealt with effectively in isolation from the other. People, together with technology, make the printing company work.

The introduction of new technology alone is only one form of potential improvement in production. The issue that new technology affects most directly is that of people, their skills, movements and knowledge. New technology may also affect people who are not directly involved in its operation, the organization around them, and possibly their relationships with managers. To achieve production potential, new technology should be supported by careful staff selection, adequate training and, above all, good communications.

WASTE CONTROL

Investment without return is a waste of resources. Potential waste areas in production are as follows:

- Ill defined levels of responsibility will probably result in delayed production and staff frustration.
- The shifting of management authority arises from weak management. Mistakes and production delays are likely to occur. Staff frustration and materials wastage can occur.
- A lack of interdepartmental cooperation will result in wasted time, effort and, most importantly, expensive materials.

Human resources:

- Inadequate supervision will lead to downtime and waste.
- Inadequate instructions will also lead to downtime, and possibly production mistakes.
- Insufficient training and development for staff will mean that the company will only achieve part of its potential.

Productive hours:

- Waiting for equipment, materials and instructions should be overcome by effective planning and production control.
- Attendance and punctuality have significant implications for production, not only for the production of the individual concerned, but also for the whole company and/or specialist departments.
- Fatigue acts as a discounting factor, and it is counterproductive. Careful consideration should be given to overall work hours.

IMPACT OF PRODUCTION DECISIONS ON PROFITABILITY

Every production decision made, whether it is good or bad, has an effect on company profitability. Also, for every way of achieving a job profitably, there are many ways of achieving it unprofitably. The provision of adequate instructions in the form of work technical specifications given to production departments is crucial in preventing waste. For each process or operation in print production, there are opportunities for error. The objective in preparing technical specifications (works/job sheets or wallets) is to distil, from a mass of detail, intelligent working instructions for departments, and to communicate them in a manner that reduces to the minimum the possibility of error.

Preventative-maintenance programmes that are properly initiated and supervised will improve efficiency considerably. Perversely, technology seems to always fail while it is producing the most urgent work, resulting in work being transferred to other units of technology, or subcontracted out. Either solution results in additional costs.

Misunderstanding some of the key factors in print production can lead to loss of time and additional expense. These factors include the nature of the work, the difficulty, line drawings, tones, screens, colours, self-colours, combinations, lamination (paper or card), (type and surface) quality, quantity, matching colours, finishing operations, the submission of proofs, authors' corrections, alterations in press, overmatter, cancelled matter, the delivery address, the bulk, the weight, and the quality relative to use. These are just some of the issues for which great care is required before and during the production process. In view of this, it is important to reemphasize the importance of:

- the selection of the right production methods and technology,
- developing the right kind of skills in production staff,
- ensuring efficient customer liaison,
- ensuring interdepartmental cooperation,
- achieving efficient communications throughout the company,
- ensuring the correct level and style of supervision (this is very important, and it is recognized that the most efficient system of supervision in production is that in which staff are encouraged rather than driven).

- Production decisions have a considerable impact on profitability, and thus planning information must be factual.
- An efficient plant layout will have a considerable impact on the improvement of productivity and profit.
- Production control that ensures the achievement of quality standards is critical to company progress.
- Quality is an attitude of mind, and it should be developed and encouraged in all staff.
- Rapid transition to the use of management information systems is occurring within the printing industry.

ACTIVITIES

- *Activity 6.1*: Identify the advantages of a method and process study to a printing company experiencing production problems.
- *Activity 6.2*: Explain the predicted results of incorrect customer liaison.
- *Activity 6.3*: Outline the main steps that should be taken to achieve and maintain production and job quality.
- *Activity 6.4*: Identify an issue related to printing production management that is not covered in this chapter.

REFERENCE

1 **Birkenshaw, J W** *Pira Guide to Management Information Systems for Printers* Pira International, UK (1991)

7

Managing printers' materials resources

Materials management entails the planning, selection, purchasing, inventory control, handling, storage and distribution of all the materials used in the printing production process. Successful management requires a sound technical knowledge combined with commercial ability, as the manager has a dual role. First, he/she must purchase materials to meet the scheduled use dates, and develop supplier relationships as part of the purchasing strategy. Second, he/she must purchase and store materials at the lowest cost that is commensurate with quality.

Materials control is a coordinating function in the procurement and distribution of materials to relevant areas or processes, and the control of all the stock retained for future allocation. Effective production depends not only on human resources and technology, but also on all the materials used in the printing production processes.

Highly trained staff and the latest technology will have little effect if the correct materials are not in the correct place at the right time and, of course, the right price. If the materials purchased prove to be unsuitable for scheduled use and the product's end purpose, financial loss and client/printer frustration will occur. The method of materials management used for the control of all the printing materials can have a great effect on productivity and profitability.

Every act, every decision, every deliberation of management has, as its first dimension, an economic dimension. (Peter F Drucker, 1982.)

MATERIALS-MANAGEMENT OVERVIEW

No one system of materials management is correct. Circumstances vary from company to company, depending on:

- the size of the company,
- any specialized production processes,
- the centralization or otherwise of the purchasing and control function,
- trade customs and practices,
- the management or other information systems in use.

Printing companies mainly sell what they produce directly to the client, and therefore most of the materials purchased are materials used in the printing production process. Some of the materials purchased are seen by the customer. These include:

- paper and other substrates,
- inks,
- binding materials, e.g. adhesives, wire stitching and book cloths,
- packing and presentation.

Many other materials used in print production are not necessarily seen by the customer. These include:

- designs and layouts,
- computer discs and tapes,
- photographic bromides and films,
- assembled foils,
- printing plates,
- processing chemicals.

These materials are often complex, and perform vital functions in the various production operations. Therefore, care should be taken to ensure that these indirect materials are not purchased on the basis of price considerations alone. While purchase prices should be as low as possible, the fitness for the end purpose should be taken into account. Cheap materials, no matter how attractive, may well prove to be a false economy, resulting in extended production times, or delayed production while the replacement of inferior materials is awaited. Thus the true skills of materials management consist in:

- matching the right materials to the job,
- obtaining materials at the most favourable price that is consistent with the required quality,
- analysing suppliers to identify specialization that confers economic advantage,
- rating suppliers for reliability,
- liaising effectively with all other departments and clients, as required,
- assessing the costs involved in holding and distributing materials,
- assessing the costs that may be incurred if some materials are not held in stock,
- maintaining accurate records.

The materials management and purchase function are shown in Figure 7.1.

MATERIALS-MANAGEMENT DUTIES

The prime duty of materials management is to analyse individual and collective job requirements in terms of all the materials to be used in the production processes, allocate specifications to the materials required, and decide whether the item(s) are to be taken from stock, or purchased individually or collectively. To achieve this, four questions must be asked:

tion with and the agreement of the customer and the printer's estimation department.

PURCHASING FUNCTION

The process of purchasing materials begins with an analysis of needs, and consideration of:

- the specification,
- the quantity and quality,
- the delivery date,
- the price,
- the order reference number.

Materials-requirements planning is an activity that is related to the planning, scheduling and stockholding of all the materials required to achieve and maintain production. In the absence of just-in-time purchasing methods, it is necessary to determine the stock safety levels that will maintain production. A stock level for an individual item is fixed by evaluating the quantity used in a given time, and making an allowance for possible unforeseen demand. When the stock falls to a predetermined fixed level, a fresh order is placed. These orders may be for fixed or variable amounts. A periodic-review order system is sometimes used, in which materials are ordered at fixed, and usually regular, intervals.

Factors that must be taken into account when stock levels and purchase are planned include the following:

- *Minimum stock*: This should take into account any possible delays during the lead time.
- *Lead time*: This is the period of time between the decision to order and when the materials are available for use.
- *Progress level*: This is the level at which enquiries should be made to the supplier to confirm the delivery date.

- *Maximum stock*: This is the stock level that should not be exceeded.

The responsibilities outlined in the section on materials-management duties can be further expanded from the purchasing viewpoint. They consist in the following:

- Achieving and maintaining reliability of supply and delivery. This point is crucial in relation to tight-schedule work.
- Maintaining adequate stock levels with the required quality. Any deviation from the quality standard will almost certainly result in customer complaint, and, in some cases, rejection of work.
- Evaluating suppliers, and their influence on the printer's ability to keep stock levels at as low a level as possible that is consistent with efficient and uninterrupted production. High stocks of materials represent committed finance that could probably be put to better use.
- Establishing supplier relations and determining specialities. Various suppliers will offer specialist ranges of materials that are not easily obtained from other suppliers. Satisfactory supplier–printer relations depend on attitudes, knowledge, and, very importantly, training in materials-management skills within the printing company.
- Evaluating the total materials costs in the long-term, in preference to obtaining low prices at any given opportunity.
- Standardizing materials as far as is practically possible.
- Evaluating and selecting new or improved materials. Care should be exercised so that any advantage offered by new materials is gained, and so that the printer avoids being used as an unwitting field tester of unproved materials.

MATERIALS-MANAGEMENT LINKS WITH OTHER AREAS

Materials management has direct and important links with all the other production and administrative functions. The origination department will require photographic films, paper, plates, proofing materials, and chemicals. The press department will require materials for production, such as press chemicals, blankets and inks, and the finishing department will require adhesives, wire, and other finishing materials. All these materials are as vital as the paper and other substrates used in production.

The purchasing function has more links with and implications for other departments than any other department. The materials management and purchasing function should be involved in continual research to establish the most favourable means of materials acquisition, and, through negotiation with various sources of supply, it should obtain the best prices for the company that are available at the time of the order.

An efficient materials-management system will develop a wide base of expert knowledge that can be beneficial for the client and the printing company. The production-planning department can benefit because of precise knowledge of materials availability, which assists in the loading of production operations.

If a system of materials quality control is operated, there will be a tighter control of materials in terms of their suitability for their purpose. This alone can result in less wastage during production of time and the materials used, and ensure that job specifications are met.

If purchasing is centralized, that is, passed through one ordering point, or decentralized, through a number of ordering points, care must be taken to identify individual department's views and requests, as sudden changes in supplier may result in considerable waste resulting from the use of unfamiliar materials. The advantages of centralized purchasing are as follows:

- When materials are ordered collectively, this can often result in useful discounts.
- There can be a consistent materials-purchasing policy.
- Moves can be made towards materials standardization.
- There can be a reduction in administrative load.
- There can be increased control over expenditure in relation to budgets.

The disadvantages of centralized materials purchasing are as follows:

- There is less control of materials selection and purchase by individual departments.
- Individual department's preferences are often ignored.
- Overall costs are likely to increase, and there will probably be less control over slippage.

The materials-purchasing function has direct links with the marketing and sales function, as any customer comments about the overall job quality will almost certainly be made to the sales people. Attention to this important link should lead to improved customer relations and satisfaction.

Materials management and purchasing has a strong impact on sales forecasts and financial administration, as the purchasing function is almost certainly the biggest spender of finance. Customer credit terms, in relation to length of time and amount, will have implications for the purchase of all directly charged materials.

INSPECTION AND QUALITY CONTROL

Quality is an attitude of mind. Advanced technology and pieces of quality-testing equipment will only function as well as human resources and calibration methods allow. However, it is important to remember that the difference between a successful printer and an unsuccessful printer may be that the successful printer has given serious consideration to quality standards that are related to customer requirements.

Although materials quality will be controlled within the company, the emphasis should be on quality at the source of supply, e.g. the supplier's performance in relation to the material's fitness for its purpose, delivery, and after-sales service. Certificates of analysis that show materials' test results can be requested. These are often obtained from suppliers of reel paper. Statistical-sampling techniques can also be used for all materials deliveries to assess the material's ability to perform without failure, losses of product quality, and increases in price because of production problems.

Testing the quality of some printing materials is difficult. An example is a delivery of paper, wrapped and strapped on pallets. In this case, quality inspection must be limited to a random check of some individual sheets, an inspection of wrapping-paper edges etc. Physical inspection of every sheet is impossible because of time and cost considerations. Problems with the quality may only manifest themselves once production has commenced. In this case, immediate decisions based on technical knowledge and experience are required. However, much valuable information can be obtained by staff from a visual inspection of stacks of paper to assess flatness, etc.

Other materials, such as film and plates, can be tested in various ways to determine quality prior to production. Any time spent evaluating or testing a material's fitness for its purpose will be repaid in reduced downtime and job wastage. It is most important that the management provide a clear definition of quality standards, and that staff are educated in the given standard. The issue of education and training is very important, for, although quality can be obtained by enforcement, this will often prove counterproductive in the final analysis, and staff may not trust a system that has been forced on them. Education about quality is a process of enabling staff to understand what their own contributions to quality are, and the importance of these contributions. Quality in the context of materials purchase means reconciling specifications with budget prices. Often, in print production, a printer has no other option but to use materials of a lower grade than he/she would prefer, providing the lower specification still meets the job specification. The professional obligation in these circumstances is to achieve the best results possible with the materials available. However, it is often a false economy for a printer to extend this policy to materials used in production, such as films, plates, inks and press blankets. Considerable care must be exercised, as low prices are not invariably associated with poor materials, but the use of 'bargain' materials often proves very expensive when their use is evaluated during production operations. Obviously, the situation depends on the type of work, the technology and the quality standard required. Establishing the correct quality for printed work is not straightforward, as quality is relative to the existing circumstances. It can probably best be defined as the standard of work which the customer will accept. This does not imply that quality should only reach the standard required by the customer. There should be a consensus of opinion about the very best quality that can be obtained with the materials available, without increases in costs, and this level should be aimed for. Any evaluation relating to materials quality control should be based on economic grounds. Efficient control results in:

- improved customer satisfaction,
- less production downtime,
- improved staff morale,
- less materials waste,
- a reduction in the number of subcontracted-production problems,
- increased efficiency in the use of technology and other equipment,
- an improved use of human resources.

Materials management and control staff in most printing companies will be unable to assess the quality of materials purchased themselves. The fitness-for-purpose tests will almost always be carried out in conjunction with the various production departments. For example the paper warehouse will inspect deliveries for apparent damage, and possibly carry out random sample assessments. The origination department will test film, plates and other materials, also by random sampling.

The printing department can only evaluate press chemicals, blankets, inks etc. through use. Although tests may be carried out, the final analysis will be based on the behaviour and use of materials during production. This is also true of finishing operations. Therefore, it is necessary for the printing company to decide:

- what quality inspection of materials is necessary,
- what materials specifications are to be established and maintained,
- what technology or equipment is required to achieve the desired quality,
- to what extent materials standardization is possible; there are distinct advantages to the standardization of materials wherever possible:
    - storage and identification become easier,
    - quality standards are easier to maintain,
    - less experimentation is required to determine the optimum use of the materials, and thus lower wastage can be expected,
    - fewer materials need to be stored.

Particular care must be taken when printed work is subcontracted and delivered directly by the subcontractor to the customer. The subcontractor quality-control procedure relating to materials supplied by him/her, and the handling of materials subcontracted to him/her should be carefully examined. A clear understanding of the quality required should be established.

MATERIALS STORAGE AND BUILDINGS

Printing-company stores vary in size and function, depending on the size of the company. However, many printers have three storage areas, which may be located together or entirely separately:

- *Production store*: This accommodates all the materials that are used in the production of the printed work, but are not sold to the customer, e.g. processing chemicals and printing blankets.
- *Inflammable-goods store.*
- *Direct-materials store*: This store contains all the materials that are directly sold to the customer, or that await allocation to a particular customer, for example paper, board and other substrates.

The direct-materials area may be further divided into preproduction storage and postproduction storage. Preproduction stores accommodate all the materials awaiting production, whether they have already been sold to a customer or are held in stock awaiting allocation. Postproduction storage may be used to accommodate work awaiting delivery or part-delivered orders. In this case, there should be careful segregation of the materials awaiting production and those in work that has been completed and stored, even when they are stored in the same area. This is particularly important when materials are preprinted and then stored to be used at a later date, for example numbered receipt and order pads, folders, and covers. In the case of security printed material, such as scratch cards and tokens, a separate and caged area may be necessary. Printers' warehouse and materials storage for small-to-medium companies is often situated within or very close to the production areas. In larger companies, storage can be in detached buildings, which may be purpose-built. The great majority of printers do not have air conditioned and temperature-controlled storage areas, and their conditions of materials

PAPER STOCK RECORD

Description ______________________ Location ______________________

Colour ______________________ Weight ______________________

Quantity ordered	Quantity Received	Quantity Issued to	Balance	Date

Figure 7.3 Paper stock record

- to enable accurate values of all the materials, both those to be retained in stock for future use and those directly allocated to a customer, to be recorded,
- to identify any loss or other misuse of materials.

Figure 7.3 shows an example of a manual materials record (similar records with major adjustments can be used for all types of materials). However, efficient management information systems are available that not only reduce operational time, but can analyse the overall situation.

MATERIALS-HANDLING EQUIPMENT

The types of equipment required to handle materials movement from stock to production and postproduction vary according to the size of the company and the type of work undertaken. Types of equipment are:

- small hand-operated hydraulic trucks that are capable of handling loads of up to approximately 100 kg,
- motorized hydraulic trucks that are capable of handling loads of up to 2–3 t, and that are possibly battery-operated,
- specialized hydraulic fork-lift trucks that are capable of lifting loads up to a height of 10–15 m, and that are usually petrol-driven.

Most of the materials-handling equipment operated by printers is capable of working in a confined space. The equipment is narrow, and capable of negotiating tight corners. Conveyors are only used in some specialized printing operations, such as web printing. These systems can include automated paper-reel selection by automated guided vehicles, delivery to individual reelstands, and the conveying of finished copies to the publishing department. Whatever materials-handling equipment is used, great care must be taken to handle all the materials in a way that avoids damage.

TRAINING

Printing materials are generally very expensive. Correct handling and control can only be carried out by correctly trained staff. Printers who readily invest in training for production staff sometimes doubt the value of training for materials-purchasing and stores staff. Correct training for staff is an investment in waste reduction, improved quality and higher profits. Training programmes should cover:

- *materials technology*: suitability, fitness for purpose, and problems associated with specific production materials,
- *deterioration prevention*: effects of temperature and humidity changes, shelf life, damage by movement,
- *handling and storage techniques*: correct movement, safety, location,
- *security techniques*: methods to guard against the unauthorized issue or theft of preproduction and postproduction materials,
- *manual and electronic recording systems*: correct use of paper records and management information systems (information on the application of MISs to materials purchase and control can be found in Reference 1)

STOCK AND INVENTORY CONTROL

The determination of suitable printing-materials stock levels is often difficult, as demand is largely influenced by the market. The great majority of printing work is individually specified by customers, and the customers have a wide range of materials to choose from. One only has to examine a paper sample book, or a copy of *Paper Facts and Figures*[2], to see the range available. This situation also applies to many other materials, such as inks, film, press blankets, adhesives and production chemicals. Obviously, no printer can store all these variations, and he/she will therefore concentrate on basic standard items, and order all the other materials in accordance with demand.

Some materials used in production can be assessed in terms of their necessary storage levels, for example the quantities of film and chemicals that are required to maintain production over a given period. An examination of materials records for the past three months and sales and marketing forecasts will show trends in usage, and allow questions about the following to be answered:

- the necessary frequency of ordering,
- the quantity of materials that must be held in stock at any one time,
- the suppliers' lead times (in the majority of cases, these are quite short).

Just-in-time ordering is often used. The JIT system requires zero, or an absolute minimum of, stocks to be held at any time. Obviously, this method of materials acquisition and supply has the advantages of reducing the finance invested in material stocks, saving space, and possibly reducing documentation. However, precise materials control from the time of ordering to the completion of work is essential in this system, as little, if any, buffer stock is available to cover unforeseen emergencies that arise during production. Considerable work is also necessary to develop good supplier relationships, and very strict quality systems at the suppliers to assure production and delivery.

A comparison of the ordered amounts, the recorded materials usage, and the quantities remaining in stock will provide information about slippage or waste. It is also possible to identify the value of the materials held, and consequently the amount of finance that is tied up that cannot be used for any other purpose.

The accuracy of materials control can be influenced by an overestimate of the waste required for any individual job. Print production incurs various amounts of waste, depending on the complexity of production. Excess materials ordered but not used or returned to store can prove very expensive.

Materials waste can also arise from:

- spoilage due to production-worker skill deficiency,
- the exposure of materials to extremes of weather,
- improper storage, with materials being badly stacked, or stored at the incorrect location,
- poor communications between the materials-control section and the production areas,
- corrected orders.

WASTE MATERIALS: PREVENTION AND DISPOSAL

It is much easier to save a pound than earn it. (Henry Ford)

Printing production operations create some natural and expected waste. The production areas that are mainly affected by waste are the printing-press room and the finishing department. Materials waste must be controlled as far as possible. Often, production workers are not aware of the costs of the materials that they use every day, and they therefore have a perception of waste that is different from that of the materials-management staff. All printing production workers should receive training in waste reduction, and be educated about the costs of the materials that they are handling. Research into the printing company's existing production methods and techniques should be carried out, and the results compared with the latest techniques. The rapid advance of electronically controlled operation is assisting with waste reduction. Better (formal) quality systems (see, for example, *BS 5750*[3]) should also help in the reduction of waste. Press-room waste usually consists of printed materials arising from press makeready operations. This waste has little, if any, value. Usually, it is sacked or baled to await collection. In some major production operations, such as web-offset printing, waste may be delivered directly to large containers, thus reducing time and cost in handling. Although this waste can be sold, the price is low, and it depends on the volume and the market conditions. Cloths used for cleaning machinery and blankets should be isolated in metal bins. Care should be taken to isolate clean and dirty cloths used in the production areas, as contamination can easily occur through close contact. Normally, these cloths are cleaned and reused. This is usually done by suppliers who specialize in the needs of the printing industry.

Finishing waste arises from:

- spoilage during production,
- the trimming of clean paper and printed work.

Clean paper shavings can be sold if they are kept separate from printed waste.

Obviously, the separation process requires time, and, although a higher price is realized for clean waste, a judgment has to be made about the tradeoff between the price obtained and the cost involved in separation. The overall influencing factors will, of course, be the volume of any one type of waste, and its market price.

The disposal of waste materials such as processing and cleaning chemicals is now the subject of health and safety legislation. (see Chapter 8: Law and the printer).

PACKING AND DESPATCH OF MATERIALS

Quality is as important in the packing and despatch of finished materials as it is in the initial selection and purchasing operation. High-quality work achieved through expert production will be a failure if it is damaged or delivered late. Also, work delivered as if it is a sack of potatoes will do little to impress a customer. Factors influencing the delivery requirements are:

- the weight, bulk and volume of the delivery (weight/volume ratios vary considerably, and they affect the decision on single or split deliveries, and, of course, the capacity of the delivery vehicles used),
- the length of the journey (immediate locality, national, international),
- the type of journey (land, sea, air).

Depending on the type of work, the packaging may consist in wrapping finished work in parcels, bulk packing on pallets, or packing in boxes. Whatever method of packing and despatch is used, close liaison with the customer is essential. Also, close cooperation with the sales and marketing staff is required to ensure that delivery dates are clearly understood. There is little to be gained from achieving a local customer's delivery date, only to discover that the work should have been delivered many miles away.

Printers often have their own delivery vehicles, or they may lease transport for the company's use, or subcontract deliveries to a haulage contractor. Obviously, the choice depends on factors such as the frequency of deliveries, distances, weights and volumes.

Some important points that must be considered are as follows:

- *Self-owned transport*: The main advantage of this is total control over the vehicle and driver. Maintenance costs can be high. Economical return loads are difficult to arrange.
- *Leasing*: The main advantage of this is control over the vehicle and staff. Under some leasing terms, maintenance costs can be avoided. Costs are usually easier to identify, as they are fixed at the beginning of the lease.
- *Subcontracting to a haulage company*: The advantage of this is that it allows the selection of the most suitable vehicle for the delivery's weight and bulk. The printer has no worries about arranging an economical return load, although this will have been considered in fixing of the haulage compa-

ny's price. Maintenance costs are the responsibility of the haulage company. Some disadvantages exist. Subcontracting to a haulage company is expensive, and the printer often has little, if any, control over the vehicle and driver.

SUMMARY

- Printing-materials management is a vital function in every printing company.
- Successful management requires technical knowledge and business ability.
- Materials control is a coordinating function in the procurement and distribution of materials to the relevant production operations.
- Modern technology and highly trained staff cannot function if materials are not in the right place at the right time.
- Nearly every decision made by materials-management staff has an economic basis.
- Materials-management systems vary according to the size of the company and the nature of its work.
- Not all the materials used in print production are seen by the customer. However the materials that are not seen are as important as those that are seen.
- It is unwise to purchase materials on the basis of price considerations alone.
- The fitness of materials for their purpose is a primary consideration.
- Materials-management duties are very important, as they have implications for the duties of all the other departments.
- The purchasing-function duty is to obtain materials at the most favourable price that is consistent with the quality required.
- Efficient material control results in waste reduction and improved profits.

8
Law and the printer

The implications of law for the printing industry are wide-ranging, and they have considerable importance for the printing manager, who is not only responsible to his/her company and the customer, but also in law.* Two main divisions of law affect the printing manager: civil law, and criminal law.

CIVIL LAW

Civil law is mainly concerned with the resolution of disputes between individuals or companies, and the awarding of compensation to wronged parties where appropriate. The great majority of print-related litigation is undertaken under civil law. An example is when a customer disputes issues of quality, price or conditions of supply. Alternatively, a printer may have to resort to law to recover money owed for work when payment is refused by the customer, or when the costs of production operations or materials are disputed.

Under civil law, an action is initiated by an individual or a company that has the right to determine how far to proceed at any one time, or to withdraw completely. A printer should remember that, under civil law, a plaintiff will succeed in cases where his/her evidence is only marginally stronger than the printer's. For the printer, the implications are clear: all impor-

* The reader should note that the matters discussed in this chapter are not exhaustive and professional advice should be obtained if problems of a similar nature to those described should arise.

tant decisions should be recorded in a written form that indicates understanding and intent.

CRIMINAL LAW

Criminal law has the function of responding to and punishing action by individuals that is regarded as objectionable to the point at which penalties should be imposed. Punishment consists of fines or imprisonment, or, in serious cases, both. Obviously, a printer involved in forging banknotes or a wide range of security documents will be prosecuted under criminal law. Action under criminal law is usually instigated by the police, and it cannot be stopped by the accused individual.

In certain cases brought under the *Health and Safety at Work Act*, action may be initiated under civil and criminal law. An example is when machinery is inadequately guarded and a worker is consequently injured. In this case, proceedings for inadequate guarding of the machine may be brought under criminal law, and proceedings for injury to the worker under civil law.

The purpose of this chapter is to deal with the legal issues that are most important to the printing manager operating in the UK. These are, in the UK:

- the responsibilities of directors,
- the law of contract,
- the law of tort,
- employment law,
- the *Health and Safety at Work Act*.

RESPONSIBILITIES OF DIRECTORS

All directors have a responsibility to ensure that the company operates according to the law. Directors' tenures of office and scope of authority are subject to the company's Memorandum and Articles of Association, which are subject to company law. A company's Memorandum of Association defines the company's objectives and the distribution of authority, and its Articles of Association list the rules and regulations for the internal operation of the company. Legal requirements for directors include the following:

- They must ensure that all the operations carried out by the company are within the law, i.e. that they comply, for example, with company law, the *Health and Safety at Work Act*, financial and authority planning acts, and employment law.
- They must ensure that the provisions of the *Companies Act* are complied with.
- They must ensure that liability to third parties is covered.

The company has an obligation to hold meetings at various times to enable business transactions to be carried out. An annual general meeting must be held every year (unless the company has passed relevant resolutions to dispense with this requirement). This meeting will usually consider the following:

- the accounts and the balance sheet, and the accountants' and directors' reports,
- the distribution of dividends,
- directors' appointments,
- the auditors' remuneration.

LAW OF CONTRACT

The nature of the contracts entered into by printers varies widely. Some involve a few pounds only, and others involve many thousands of pounds. The great majority of contracts relate to the supply of printed and similar materials, and they are subject to various statutes which govern the interpretation of contracts. In cases of contractual dispute, it is wise to seek advice from a solicitor, as the whole issue of the formation and performance of contracts is complex. A printing company should not enter into action without expert consideration of the issues involved in each individual case.

Formation of contract

Contracts are based on the acceptance of offers or orders, in which details of the agreement to provide printed work, and the consideration (or payment in money or money's worth), are given. Some 'contracts' may be enforceable without any 'consideration' if made under seal. However, such contracts are not discussed in this chapter. The contract will usually contain details of technical specifications, price, delivery times etc.

However, contracts do not have to be in writing for a contract to be binding on the parties. A contract exists where verbal agreement has been reached between the parties, provided that certain essential requirements are fulfilled, which include:

- A clear offer is made and accepted (whether by deed or otherwise).
- The parties involved intend the contract to be legally binding (an agreement to agree terms at some future date does not form a binding contract).
- Each party commits to providing the other with something and receiving something in return (e.g. services for remuneration).
- The terms and conditions of the contract must be capable of legal enforcement. A contract to commit any illegal act would not be enforceable, on the basis that it would offend against public policy.

Misrepresentation may affect the validity of a contract. Misrepresentation can be defined as a false statement made by one party of the contract to the other which induces the other to enter into the contract. There are three kinds of misrepresentation: fraudulent, negligent and wholly innocent. The remedies for misrepresentation vary, depending on the state of mind of the person making the misrepresentation. For example, should the misrepresentation be fraudulent (i.e. made knowingly, without belief in its truth, or recklessly, without care being taken to determine whether it is true or false), the other party may terminate the contract and may make a claim for damages.

An example of misrepresentation would be when, during negotiations between printer and customer, the customer asks if all the printing production operations will be carried out and completed on the printer's premises, and the printer affirms this to be the case, and subsequently subcontracts the work. In this case, misrepresentation has occurred, and the customer may terminate the contract and claim damages.

Performance and breach of contract

The performance of contract is covered by a general rule that is quite simple: both parties to the contract must completely and precisely perform what has been agreed. Failure to perform the contract completely, means that a contracting party cannot sue on the contract, but can be sued.

The exceptions to the general law (referred to above) are as important as the general rule itself:

- *Severable contracts*: Whether any particular contract is severable depends on the intention of the parties at the time of contract. If a contract is severable, part performance may allow the injured party to have the right to sue for damages, but not to terminate the whole contract.
- *Prevention of performance*: If a party to a contract performs part of the work he/she has undertaken, and is then prevented by the fault of the other party from completing the work, he/she can sue the other party for damages.
- *Acceptance of partial performance*: If a party accepts only partial performance from the other party, it may be said that the parties have varied the initial contract, and the

- Conditions about the use of the customer's own materials should be specified. Although customers' materials will be protected from damage as far as possible, any claim relating to damage by fire, theft or accidental damage cannot exceed the amount recoverable from insurance.
- Production schedules should be agreed by the customer and the printer. However, neither should incur penalty or liability for delays due to reasons beyond their control, such as government action or war.
- It is most important that delivery conditions and prices be clearly understood. Prices and conditions will obviously vary according to circumstances. However, in the majority of cases, there will be a single delivery to the customer's premises or other address stated in writing. The customer should understand that, unless it is stipulated in writing, the price quoted does not include deliveries to the customer's clients. The ownership of the work will be transferred to the customer at the time of delivery or at time of invoicing, whichever occurs first.
- The terms of payment are an important part of any printing contract, and they depend on the relationship of the printer and customer. Payment against a *pro forma* invoice may be required in some cases. In others, extended credit may be given. Normally, net payment is required 30 days from delivery.
- It is wise for a printer to seek indemnification against claims resulting from infringement of copyright.
- Printers are not required to print any matter when this would be, in their opinion, illegal.
- A printer would normally expect any claims relating to work to be made within ten days of delivery. Claims from the customer can present particular problems for the printer. A customer cannot reasonably be expected to inspect every single copy delivered. A random inspection only may be carried out, and therefore some length of time may elapse before it becomes obvious that there is faulty work. Work may have to be reprinted, or financial compensation may

have to be made to the customer. These problems will be reduced to a minimum if efficient quality-control systems are operated at each stage of production.

The British Printing Industries Federation and other organizations throughout the world produce versions of conditions of trading for the industry. Some companies may prefer to prepare their own conditions. It is important to understand that the printing of conditions on the back of an estimate or quotation is not enough. The customer's attention should be drawn to these by annotation on the face of the estimate or quotation. It is also important to recognize that any condition which is included in any standard terms of trading may be void or subject to the court's interpretation as to its reasonableness (or otherwise), depending on the nature of the contract and who the parties to the contract are. When any standard terms are created, it is advisable to take legal advice on the effectiveness of the terms that it is proposed be contained in such conditions.

COPYRIGHT

Copyright is a property right which a person has in relation to the work that he/she has created. He/she therefore has a right to stop others from copying or exploiting the work without his/her consent.

Therefore, printers who reproduce copyright work without the owner's consent obviously infringe copyright. Provided that, at the time of issuing legal proceedings, the plaintiff has title to the copyright, which is an essential ingredient of the cause of action, the plaintiff does not have to prove that he/she has suffered any damage to succeed.

The *Copyright, Designs and Patents Act 1988* gives the legal details and requirements of copyright law. This act is a complicated piece of legislation, which creates separate rights which may coexist at any one time in relation to a piece of work or subject matter. The determination of who owns a copyright, and is therefore entitled to give permission to reproduce, can be difficult. In all cases, a printer would be prudent to insist on written confirmation from the customer that the obligations of copyright law have been complied with. The printing company should seek the appropriate licence or consent prior to undertaking any printing/publication work.

OFFENCES

Under the *Obscene Publications Acts 1959 and 1964* penalties may be imposed on the printer for various actions. These include:

- printing and publishing indecent matter,
- sending obscene or indecent printed matter through the post,
- printing and publishing proposals for lotteries,
- committing libel through the printing and publication of defamatory statements.

IMPRINTS

There are statutory requirements under the *Newspaper, Printers and Reading Rooms Act 1869* about the use of a printer's imprint for certain printed matter. Any company printing papers or books for publication and distribution must include the company's name and address on the printed work. Certain items are excluded:

- banknotes,
- receipts for money,
- business cards,
- price lists,
- reproductions of engravings,
- parliamentary documents,
- dividend warrants and transactions.

Menus and catalogues are often printed without an imprint, and the printer is therefore open to prosecution, although this is very unlikely.

A difficulty arises with printers' imprints when a printer subcontracts printing operations. The subcontracted printer should show his/her imprint. However, this may cause considerable embarrassment to the first printer or his/her customer. Printing-trade custom does not require the subcontractor's imprint. Unfortunately, this will not provide a defence if a prosecution is initiated. A customer's request that a printer's imprint should be omitted does not absolve the printer from imprint obligations. If the law requires an imprint for a particular job, the printer is under an obligation to insert it, and, at the same time, point out the law to the customer. It is important to note that the address must be that of the printer, and no one else.

An imprint on work will also assist the printer in disputes about issues of production. Therefore, all the proofs of work should include the printer's imprint at the outset, and this should not be left out until the final stage before printing.

LAW OF TORT

As the law has developed, it has been recognized that certain relationships give rise to a duty of care. There need be no contractual connection between the parties to these relationships. If a duty of care is owed, then a breach of the duty which causes damage to the other party will entitle the other party to bring an action for damages.

In addition to the general duty of care, several statutes have been incorporated into English law which create statutory duties. These include the following.

Occupier's liability

The Occupier's Liability Acts 1957 and 1984 state that, when a person suffers loss or is injured upon premises 'occupied' by another, the injured person may be able to sue for damages. In some cases, it may be possible to exclude the duty by contract. It is, therefore, important for printers to maintain their premises well, and to consider limiting any potential liability to 'visitors' entering onto their land.

Additionally, an employing company may be held to be liable for the acts of its employees under the rule of 'vicarious liability'. The general rule is that, if an employee commits an act in the course of his/her employment which causes loss to a third party, the employer may be sued in 'tort'. Printers (and indeed any employers) should realize that their employees may expose them to liability to third parties. It may be possible in some circumstances to insure against such risks, and it may be worthwhile discussing this with insurers.

EMPLOYMENT LAW

The selection, recruitment, employment and control of staff are vital issues for the printing manager. These issues are subject to various pieces of legislation which together can be termed employment law.

The majority of issues in this area, such as contracts of employment, dismissal, and the redundancy of employees are matters of civil law. Issues of safety, such as the guarding of machines, the safety of dangerous chemicals, and safe working conditions are subject to other legislation, and criminal-law statutes may be applicable.

Employee status

An employee is normally defined as one who offers skill and labour in return for a wage or salary from an employer. However, a company may employ its own directors, and, in this case, the directors are both employers and employees.

A distinction has to be made between employees and independent contractors. There are several tests which differentiate between an employee and an independent contractor. For example, a worker may be defined as an employee when he/she is controlled day-to-day by the company that uses his/her services and pays him/her. An independent contractor is not necessarily dependent on work from one source, and he/she may work without supervision. The method of paying workers may assist one to define whether the worker has employee or independent-contractor status. Employers are responsible for the administration of tax, and must make deductions from the salaries of employees (of Pay As You Earn tax and National Insurance). The independent contractor, on the other hand, carries out his/her own tax administration, and must charge the person who is paying for his/her services Value Added Tax if the independent contractor is registered for VAT.

Contracts of employment

A contract of employment can be made orally or in writing. To overcome misunderstandings or disputes that may arise from oral contracts and employment agreements, the majority of printing companies will issue written contracts of employment. However, a statement, whether contained within a written contract or otherwise, notifying the employee of certain specified terms of employment must be given to the employee within 13 weeks from the start of his/her employment, and failure to do this is an offence by the employer. A written contract of employment normally includes *inter alia*, terms that identify the:

- The identity of the parties to the contract.
- The date on which the employment began. (This is important, as certain statutory rights, such as redundancy payments and payments for unfair dismissal, depend on the length of service of the employee.)

- The scale or rate of remuneration, or the method of calculating remuneration, and the intervals at which remuneration is to be paid. Under the *Employment Protection (Consolidation) Act*, all employees have the right to an itemized pay statement in writing, either before or on the day on which payment is made. The statement must include:

  - the gross amount of wages or salary,
  - the amounts of and reasons for all deductions from the gross amount,
  - the net amount of salary payable,
  - in cases in which different parts of the net amount are paid in different ways, the amount and method of each payment, clearly shown.

- The terms and conditions relating to hours of work. Requirements for overtime or shift working should be shown, and the contract should state whether any overtime is mandatory, i.e. whether it is regarded as part of the normal spread of hours. Working hours are usually made subject to company requirements from time to time. Any provision for flexitime should be clearly indicated.
- The terms and conditions relating to holidays, or holiday pay. The terms relating to holiday time, holiday periods and holiday pay are currently more likely to be subject to individual agreement, rather than collective agreement with trade unions.
- Incapacity for work owing to sickness or injury, including any provision for sick pay. A clear statement should be made about whether payment will be made during periods of sickness, and the amount of any payment.
- Pension arrangements. The contract should state whether the company has a pension scheme, and, if it has, its status under the *Social Security and Pensions Act*.
- The length of notice of termination of employment that the employee has to give and is entitled to receive.
- The employee's job title.

- The employee's place of work. This should be widely specified.

The contract should also include a note about any grievance or disciplinary procedure used within the company, and it should specify a person to whom representation should be made. The steps involved in such proceedings should be easily understandable.

Any change in the conditions should be confirmed in writing to the employee within one month of the change.

Although they are under no legal obligation to do so, many companies will attach a job description or specification to the contract of employment. However, an employer should be careful that this does not affect his/her ability to move an employee around the business, or hinder flexibility generally. Some terms will also be implied in the contract of employment. These are implied common-law duties, and duties that are long accepted and well known in the printing industry.

General duties of employer

The prime duties of an employer are:

- the taking of reasonable care and precautions to ensure the safety of employees,
- the provision of work, although this may be overridden by an express provision in the contract,
- the indemnification of employees against liabilities and losses (the employee should be indemnified against all liabilities and losses properly incurred during the cause of his/her work),
- the provision of a safe system of work, e.g. the provision of safe premises, technology, tools and equipment,
- reasonable behaviour (employers owe an implied contractual duty to behave reasonably and responsibly towards their employees, and the implied duties are mutual respect and consideration).

Duties of employee

The prime duties of an employee are:

- *Obedience*: There must be cooperation with the employer to achieve lawful company aims and objectives. Employees are not obliged to carry out any instruction outside their contract of employment. An employee can either ignore an instruction given outside the terms of the employment contract, or, if the instruction involves a fundamental breach of contract, the employee can regard it as constructive dismissal.
- *Fidelity*: Employees must take care not to place themselves in a position in which their own interests conflict with duties owed to the employer. The employee must:
  - not compete with the employer,
  - not disclose confidential information about the company or its clients, or use information to benefit him/herself; an employee will breach his/her own fidelity if he/she shows that he/she is unworthy of the trust invested in him/her.

An employee must take reasonable care when carrying out duties. Care must be taken of the property, technology, tools and equipment entrusted to him/her. Also he/she must be careful that any actions that he/she takes will not harm fellow workers or outside contractors, thus imposing vicarious liability on the employer. The employee owes a duty to produce work to the best of his/her ability at all times. Acts of gross negligence may, in some circumstances, entitle the employer to dismiss the worker without notice.

Discrimination

It is unlawful for an employer to discriminate against existing and potential employees on the grounds of sex, race, colour or ethnic origin. Therefore, particular care must be taken when preparing advertising materials, recruitment, employment and appraisal schemes, and dismissal procedures. The *Sex Discrimination Act 1986* and the *Employment Act 1989* have resolved many of the restrictions formerly imposed on the employment of women and young people (young people are, for the purposes of the Acts, those under 18 years old).

Generally, companies with 20 employees or more should ensure that 3% of their staff are people who are registered as disabled, and they must not recruit people not registered as disabled people or dismiss a registered disabled person without reasonable cause, if this would reduce the quota below the limit. However, there are exceptions to the rule, for example if the working conditions are hazardous. An employer must apply for an appropriate permit if this is the case. The majority of print production operations can be hazardous, and they are therefore regarded as unsuitable for disabled persons.

The *Equal Pay Act 1970* operates in parallel with the *Sex Discrimination Act 1986*. Under the *Equal Pay Act*, terms and conditions of employment should be no less favourable for females than they are for males doing approximately the same type of work and having the same levels of responsibility.

Statutory sick pay

Employers are liable for the payment of statutory sick pay under the *Social Security and Housing Benefits Act 1982* to their employees. Subject to contribution requirements, statutory sick pay will be paid for up to 28 weeks of absence due to sickness or injury. Part-time staff who earn above the relevant earnings limit can also claim statutory sick pay in the same way as full-time staff. The qualifying conditions for, and limitations to, entitlement are complex, and therefore the print manager should take advice or familiarize him/herself with all the relevant documents.

Sickness must be notified to the employer. However, evidence of sickness must be supported by a doctor's statement,

The preparation and publication of, and adherence to, a company disciplinary procedure will assist in the achievement of a fair termination of employment. It will also communicate to the staff the standards of behaviour that are required, and the standard of behaviour that will not be tolerated.

Redundancy

An employer has an obligation under the *Trade Union and Labour Relations Act 1992* to, in certain circumstances, notify trade-union officials and the Secretary of State for Industry in writing of impending redundancies, and consult with trade-union representatives. The information given should include:

- the reasons for the redundancy proposals,
- the number of and a description of the employees that it is proposed be made redundant,
- the number of employees, by type and grade, to be made redundant,
- the proposed method of selection for redundancy,
- the proposed ways of achieving the dismissals, and a statement of the period over which the redundancies will be spread.

Redundancies may take place for the following reasons:

- The business for which the employee was engaged is to cease.
- The business is to be moved to a different location.
- There is a reduced need for employees to carry out work of a particular type at the place at which they are employed.

Employees are only entitled to a redundancy payment if they have been continuously employed by the same company for at least two years.

Printing managers should never take the situation of staff redundancy lightly. Redundancy will have a dramatic, if not traumatic, effect on the people who lose their jobs. Every effort should be made to redeploy, retrain or relocate people where possible. Redundancies affect not only those who leave a company, but also those who stay, because of increased uncertainty about the future.

Maternity rights

Maternity rights exist and are imposed on employers in certain defined cases if employees qualify for maternity rights as laid down in English statute law.

HEALTH AND SAFETY

Health and Safety at Work Act 1974

All printing companies in the UK are subject to the legal obligations of the *Health and Safety at Work Act.* The act requires every employer to ensure, as far as possible, the health, safety and welfare of the company's employees.

The act requires

- the absence of risks to health relating to the handling and storage and movement of articles and substances, as far as is reasonably practicable,
- the provision of safe entrances to and exits from the place of work,
- the provision and maintenance of a working environment for employees that is, as far as possible, safe and without risks to health,
- the provision of printed information, instructions, training and supervision as necessary to ensure safety at work,
- the provision and maintenance of technology, equipment and tools that are necessary to complete work safely.

It should be assumed that these provisions cover clean drinking water, toilet and washing facilities, and a suitable area in which to leave and/or change clothes.

Company health and safety policy

The *Health and Safety at Work Act* requires all employers to do the following:

> prepare, and as often as may be appropriate revise, a written statement of his general policy with respect to the health and safety at work of his employees, and the organisation and arrangements for the time being in force for carrying out that policy, and to bring the statement and any revision of it to the notice of all of his employees.

The only exception to this requirement is when a company employs fewer than five people. The *Health and Safety Act* also imposes a duty on the employer to ensure the safety of people subcontracted and working on his/her premises. It is important for the manager to note that there is no option but to comply with the legislation. The duty is absolute. It is no defence to say that one did not intend to breach the regulations, or that they were breached in ignorance. An example of absolute duty is seen in the legislation relating to the guarding of machinery. All prime movers, transmission machinery and any other parts of dangerous nature must be within the regulations of the acts, which state, for example, that:

> all guards shall be of substantial construction and constantly maintained, and kept in position while the parts required to be fenced or safeguarded are in motion or use, except when any such parts are necessarily exposed for examination and for any lubrication or adjustment shown by the examination to be immediately necessary and all such conditions as may be specified in regulations made by the Minister are complied with.

Additionally, all machinery, new or second-hand, must be securely guarded in accordance with the acts before sale. Non-compliance may be regarded as a criminal offence.

Factory inspectors are appointed by the UK Government to monitor and supervise compliance with the acts. These inspectors have the right to enter any company and any part of a company at any time of the day or night. Points that will receive attention are:

- the cleanliness of the working areas,
- the accident records,
- the guarding of machinery,
- the entrances and exits,
- the fire precautions,
- compressors and lifting gear,
- general welfare provisions,
- toilet facilities,
- first-aid provision,
- provision for the storage of dangerous chemicals.

The general duties of employees relating to health and safety include the following:

- They must take reasonable care in relation to the health and safety of themselves and other persons who may be affected by their acts or omissions at work.
- They must cooperate with the employer and/or others having statutory safety obligations as much as is necessary to enable the duty or requirement to be complied with.

Control of Substances Hazardous to Health Act 1988

The *Control of Substances Hazardous to Health Act 1988* (COSHH regulations) requires a company to provide printed information about, and provide training in the use of, all hazardous substances used at work. Whereas the *Health and Safety at Work Act* covers the general environment, machinery, and ease of movement, the COSHH regulations concentrate on the controlled use of substances that can affect the health of people using them. The regulations are intended to act as a framework for the employer in the protection of people at their workplace.

Complying with the COSHH regulations entails (according to the UK Health and Safety Executive):

- assessing the risks to health arising from the work,
- deciding what precautions are needed (a company must not carry out any work which could expose employees to hazardous substances unless it has assessed the risks and devised the necessary precautions; unless the company has judged the risks correctly, it is unlikely to decide on the right precautions),
- preventing or controlling the risks (this entails assessing and evaluating the risks and establishing safe methods of dealing with them, and ensuring that control measures are used and maintained properly and that any safety procedures that have been laid down are followed),
- monitoring the exposure of workers to hazardous substances, and carrying out appropriate health surveillance where necessary,
- informing, instructing and training employees about the necessary risks and the precautions.

SUMMARY

- Both civil and criminal law must be considered by the printing manager.
- Directors and managers have duties under law that cannot be avoided.
- The majority of printed work is completed under some form of contract, and the complex rules of contract law govern such work.
- A contract does not have to be in writing to be enforceable in law.
- Standard conditions of trading constitute methods of avoiding misunderstanding about contract terms and restricting

liability. However, any restriction of liability included in standard conditions may be of no use if it is 'unreasonable' or unclear.

- Copyright law has particular significance for the printer.
- Employment law cannot be ignored by a print manager without problems being invited.
- Contracts of employment are vital employment documents that contain very important information.
- Employers and employees have obligations under the *Health and Safety at Work Act*. These duties are mandatory.
- Discrimination is, in nearly every circumstance, illegal.
- Maternity rights are enforceable in law.
- An employer has statutory obligations under the *Social Security Act*.
- The termination of employment may be fair or unfair. The distinction is very important to the print manager.
- Strict regulations apply to redundancy.
- The *Health and Safety at Work Act* is a major piece of legislation that affects all printing companies in the UK.
- An understanding of the COSHH regulations is essential for all printing managers.

ACTIVITIES

- *Activity 8.1*: Carry out a COSHH assessment for any substance used in printing to evaluate risks to health.
- *Activity 8.2*: Identify the key requirements in forming a valid contract enforceable in law.
- *Activity 8.3*: Prepare a contract of employment for a printing production worker, using any production job with which you are familiar.
- *Activity 8.4*: Identify a legal issue that may affect the printing manager that is not dealt with in this chapter.

9
Managing change

The previous chapters of this book have described many of the issues and disciplines involved in print management. These disciplines are integrated into the field of management in the printing industry. However, the only means of applying management techniques and strategies to achieve the company's business aims and objectives is through the manager him/herself.

Without efficient, innovative and vision guided management, nothing will happen. The company will not progress, and it may eventually cease to trade. Radical change presents the manager with serious problems in analysing trends and determining how they are likely to affect the company. The effective management of change depends almost entirely on the management approach to the situation, and its attitude and strategy in circumstances of change.

Two statements in Reference 1 are of very special interest to the print manager: 'there will be a growing need for skills in the management of technology and innovation', and 'the diversity of skills required in the management functions will need an entry calibre of degree level'. Phil Jesson, writing in the *British Printer*, has stated the following. 'Many British managers fail to understand or apply principles of leadership, inspiring willing action in others, and instead hide behind an autocratic impersonal style. But leadership is now even more important in industry than technology.'

Thus the challenge of modern times facing the printing industry is the development of people with management abilities to take the industry forward to the next century.

STATUS OF INDUSTRY

For many years, the printing industry enjoyed expansion, a large workforce, and sufficient work to allow many small companies to develop. Today, the industry is faced with increasing competition, not only from within the UK, but also from companies throughout the world. Customer demands for quality production are intense and at the same time prices are being subjected to greater pressure than at any time in the past. These changes have been accelerated by recession, but, even in the absence of recession, it is certain that the industry would still have been faced with the need to change, as progress can only be avoided or resisted in the short-term.

Rapid advances in production methods and technology have also resulted in a contraction of the industry's workforce. Skills requirements have completely changed and now a higher level of educational achievement is necessary for those entering the industry at all levels. For the printing manager, the day is fast approaching when degree levels of education for the potential manager will be standard.

Closely related to these requirements is the need to increase effective production and profitability. This is a difficult task for the manager with little knowledge of business or the skills requirements of disciplines outside his/her immediate experience.

Reference 1 indicates the following:

- There is currently a lower economic growth rate than there has been over the last ten years.
- There is intense competition worldwide in both publishing and print, but particularly in print.
- Legislation relating to environmental issues and the reduction of waste and pollution is increasing.
- Further legislation is being passed relating to health and safety at work.
- There is a greater emphasis on customer services than there has been in the past.

- Skills requirements will continue to change.
- Analytical skills will replace some craft skills.

MANAGEMENT ATTRIBUTES

In these circumstances, the manager becomes the key to progress, and he/she must therefore possess certain attitudes and abilities. Most important is the recognition that management is a discipline, and that it entails working with people who have differing perceptions of work and life in general. Any manager who wishes to be in an advantageous position when change affects the company needs to understand these important facts of modern management. A manager should have the following attributes:

- *Company management*: The manager should be part of the company management team, and be seen to be forward-looking. He/she should have the ability to detect trends that may require changes in the company's operation.
- *Leadership*: A manager should be a leader. Leadership is recognized as being a major influence on productivity and company success. Leadership can be defined as the ability to inspire or motivate staff to achieve company objectives. Leadership entails the recognition of staff achievements and the importance of the work contributions of members of staff.
- *Technical 'knowhow'*: The manager should have an adequate technical knowledge of methods and processes and the technology used in production. He/she need not be able personally to operate effectively each item of technology or process, but he/she must be able to control those who do.

Prior to any response, decision or strategy for change, it is necessary for managers to assess their own individual approaches to managing change. A self-diagnosis of management style, attitude and approach entails asking the following questions:

- Do I have a thorough understanding of the need and objectives for change?
- Am I aware of the implications of change upon the staff?
- What is my attitude to company change, innovation and technical opportunity?
- What is my view of staff involvement in determining and planning change?
- In what way can I communicate proposed changes to staff?

Obviously, a negative approach to any of these questions indicates a management problem that will, at the outset, prevent any satisfactory progress.

Assessments of managers and their abilities can also be carried out by senior managers. In this case, the following management attributes are apraised:

- leadership, acceptance of responsibility, positive and innovative attitudes,
- the ability to communicate with staff,
- the ability to work with others in a team,
- the ability to be assertive without being aggressive,
- the ability to recognize and resolve problems,
- integrity, to company requirements.

MANAGEMENT STYLE

Management style has a major influence on the achievement of results. Management styles can be classified as follows:

- *Autocratic*: This is a dictatorial management style. Decisions are reached without consultation. Instructions are issued and are expected to be followed without dissent.
- *Democratic*: The democratic management style encourages initiative and motivation, and believes in consultation and the delegation of authority. However, all democratic man-

agers need to resort to autocratic management under certain circumstances.

- *All-alone*: This style is practised by the manager who believes that no one else has the ability to manage the situation. Its use indicates a failure by the manager to understand people and their abilities. Little is likely to be achieved if this style is used.
- *Leadership*: The manager who uses this style operates on the basis of leading by example and involving others in management decisions.
- *Motivational*: This style is often based on the provision of incentives, either intrinsic or extrinsic.
- *Monitoring*: The manager who uses this style operates by delegating authority, and then closely supervising performance.

Combinations of these management styles are often used to deal with specific circumstances, as no one style of management will be suitable for all situations.

MANAGEMENT OF CHANGE

The management of change can be defined as the development of a strategy that allows a company to react quickly to changes that may influence the company's progress or stability. This requires managers to be forward-looking, innovative and able to detect trends within the industry that may affect overall performance. An ability to recognize and exploit opportunity is also needed. It is unlikely that an autocratic approach to change will succeed without disruption, frustration and possibly antagonism. The ideal style of management for change is one in which the management announces the needs or reasons for change and the proposed methods of change and asks the staff for comments or suggestions. Management does not abrogate its authority, but analyses responses to proposals, and considers all the points raised. Where possible, original proposals are amended. This approach:

- ensures that staff problems are known,
- encourages teamwork between the management and the staff to achieve change,
- allows the delegation of some change responsibilities,
- assists in reducing any staff fears and suspicions about the proposed changes.

Assessing need for change

All printing companies, irrespective of size, need to recognize the need for change that results from industrial and economic progress. Changes within the industry are a continuing process, and they are inevitable. However, change for change's sake can be counterproductive. Any major changes should be based on definite and compelling reasons, such as the following:

- The company experiences problems with performance, because of changes in technology, quality demands, delivery, prices, contraction in the customer base, low profitability etc.
- A change of emphasis in processes or methods is necessary that may have an impact on the customer base.
- The customers produce more of their own origination, e.g. DTP and page makeup.
- The management structure is inefficient, and frustrates aims and objectives.

HANDLING CHANGE

A method for handling change following an assessment of needs is as follows:

Explain and communicate. Straightforward communications depend on the successful transfer of information from one person to another. All communications about proposed changes should contain a clear explanation of the reasons for change, and how it may affect individual staff. This is particularly important in terms of gaining the confidence and support of staff who will either be involved in the planning process, or affected by the proposed changes.

Consultation

Consultation allows ideas to be received from staff about change and how it is to be implemented. Staff ideas are sought about decisions that will affect them, directly or indirectly. Consultation may also bring to the management's notice any important factors that have been previously overlooked. Staff will accept change more readily after consultation, as they will feel involved in the company's future, and will believe that their views are recognized and valued. Any specific worry of the management or the staff will probably become apparent, and thus be more easy to deal with.

ANALYSIS AND DEVELOPMENT OF HUMAN-RESOURCES STRATEGY

Analysis should consider the results of consultation and any subsequent amendments made to the original plans. Staff retraining, redeployment or redundancy should be dealt with at this stage.

Sometimes, an attempt is made to resolve issues by curtailing investment in technology, buildings etc. to save jobs. This course of action is unlikely to succeed, as it is rarely possible to bring about sufficient savings to resolve the situation. The result may be an overstaffed company with inferior technology. Unfortunate as it may be, if a genuine need to reduce staff is accepted, then some of the normal ways of achieving this should be attempted first. Obviously, natural wastage is one of the most popular methods. However, this can only be really effective when it is known that staff will be leaving the company in the near future. Early retirement can also be considered, but this may prove expensive if long-standing employees are involved. Natural wastage is further hindered by the employment situation within the industry. People are less likely to move jobs during a period in which staff reduction is common. It is also unlikely that natural wastage will occur in all departments in the required proportions.

Voluntary redundancy is another option. At first sight, this appears to be a satisfactory way of resolving the problem. However, experience has shown that too many staff may show an interest, especially if the company is offering enhanced financial severance offers. Also, voluntary-redundancy requests will not be received only from older staff. Younger staff may also express an interest, and thus present the manager with some difficult decisions. Not all can leave, how is the selection to be achieved? What will be the effect on those wishing to leave but being refused? It is to be hoped that these situations will become less common as print companies adjust to the manpower needs of the modern industry.

Forced redundancy should always be seen as a last resort. However, it obviously has the advantage of being quick to implement, and staff reductions can be made in specific areas as required. All managers involved in the problem of reducing staff numbers should recognize the trauma for the average employee who is told that the company has no further need for his/her services. Enforced redundancy not only affects those losing their jobs, but also other employees within the company. A feeling of instability can develop, and alter staff attitudes. An atmosphere of fear and tension may exist, and certainly morale will be damaged. Therefore, every possible attempt should be made to retain or redeploy staff.

It can be seen that there is no entirely satisfactory way of dealing with this issue, but the situation can be mitigated by effective communications.

Changes in technology and working practices create pressures on the provision of human resources that will not always result in a need to reduce staff levels. The need for recruitment may also arise. The printing industry has been subjected to scientific and technological change for some time, and this situation is not likely to alter in the foreseeable future. The result has been a progressive change in skills demands for all areas of the industry. Many of these skills will have electronic knowledge and computer literacy as essential components.

Staff recruitment during these changes provides the opportunity for the company to employ the right person for the job, through a correct assessment of the attitudes, skills and values required. The ultimate aim should be to develop a team approach to the company operation. The team approach will normally allow greater staff flexibility and increased consensus of opinion.

The process of finding and keeping the right staff can be expensive, but it should be viewed as a wise investment in the company's future. It is a fact that no company can successfully manage change and achieve success in the future without correct human-resources provision. Therefore, any opportunity or need to recruit staff should be given considerable management thought.

The key issues of staff recruitment include assessing the skills, knowledge and attitudes that are needed for the company to be able to compete, as follows:

- Decide exactly what is required for each individual job.
- Prepare job descriptions. The correct preparation of a job description will lead to a clear definition of the type of person that is required for the job.
- Prepare a person description. This will include the appropriate age range, attitudes, skills, ambitions etc.
- Clearly indicate any individual development required. This requires management to think ahead, and evaluate the possibility of additional pressures on the employee resulting from changes in technology, techniques or the environment.

PLANNING AND IMPLEMENTING CHANGE

The preparation of plans designed to change a company requires very careful consideration. The most significant problem in long-term planning is the difficulty of predicting the future. The printing industry is rapidly changing because of advances in technology, economic considerations, and customer demand. There are only three facts that relate to forecasting the future:

- Change will continue, possibly faster than ever before.
- The future will certainly be very different from the past.
- The future will always be difficult to predict.

Developing plans of action, or projects for change, to respond to expected circumstances within the industry must therefore involve some risk. The risks will be considerably reduced if a carefully thought out plan is developed and acted upon. Any plan cannot, in itself, deal with all problems and resolve all situations resulting from change, but it can minimize the degree of surprise resulting from unexpected changes.

The effectiveness of any plan will, to a large extent, be determined by the management and its approach. Aims and objectives for change can only be achieved through staff, and a participatory approach will achieve far more than authoritarian attitudes. Modern management can no longer rely entirely on power and authority to achieve required results, and it should recognize the value of staff involvement. Some managers assume that staff participation will reduce his/her management authority and control. This is only likely when the management is weak and lacks in direction.

Management competence is the key to successful change:

- Managers must have an ability to create an environment in which teamwork and individual initiative can operate and develop. This requires strong and flexible leadership.

- Managers must present a vision for the future of the company, and obtain the commitment of the staff to its achievement, by involving the staff in planning and organization. Intrinsic motivation through staff enthusiasm can be achieved by involving the staff at all the stages of planning.

Planning takes two forms:

- *Operation planning*: This is usually short-term, and it is designed to resolve minor operational problems.
- *Strategic planning*: This is usually long-term, covering a period of up to five years. Strategies are based on expected needs, and they must be matched to the strengths and weaknesses of the company.

JOB SATISFACTION THROUGH INVOLVEMENT

Job satisfaction is improved when staff become involved in the problems of human resources, technology, production processes and methods. Involvement does not mean that staff totally control the solution of these problems, but that their views and ideas are taken into account. Often ways of overcoming problems of planning or production are identified by the staff involved, but not by the management.

If any of these solutions are adopted by the management, it will receive the full support of the staff who originated the ideas. The benefits of improved job satisfaction through staff involvement in company affairs are as follows:

- The staff involved are able to show and use their capabilities, and thus they are more likely to use self-direction to improve their contribution to the aims and objectives of the company.
- The staff broaden their skills through enhanced experience, and they are probably more receptive to any proposed changes.

- The great majority of staff are capable of improved job performance. They are encouraged by their perception of their identity within the company.

There is a considerable amount of evidence in the Hawthorne reports[2], which consist of a large number of interviews with workers in various departments, to support the view that work efficiency and job satisfaction are closely linked, and that attitudes towards the company are very important factors in motivation.

RESPONDING TO CHANGE

The need for change is almost certainly to arise, and it will need to be resolved. Care should be taken to view any need for change in a positive manner, as an opportunity for improvement. An analysis of existing company performance will provide valuable information as change indicators. Analysis can be carried out using a capability profile, in which a number of issues are considered:

- the company's present situation in relation to the market, technology, human resources and financial resources,
- the company's short-term and long-term objectives,
- the management approach to any action being taken within the company.

Examples of documents used in preparing analyses and plans are as follows:

- a development plan (see Figure 9.1),
- the planning assumptions (see Figure 9.2),
- the planning objectives (see Figure 9.3),
- a SWOT analysis (see Figure 9.4).

DEVELOPMENT PLAN

Department or project ____________________ Date __________

____

Manager or team ____________________ Page ____ of ______

Programme (intended development)	Action planned (objectives)	Action standards (standards determining successful outcome)	Action outcome (result of action)	Date

Figure 9.1 Development plan

<table>
<tr><td colspan="4">PLANNING ASSUMPTIONS

Department or project ____________ Date ______________
____
______________________________ Page____of _______
Manager or team ________________</td></tr>
<tr><td>Change planned</td><td>Optimistic result</td><td>Pessimistic result</td><td>Most likely result</td></tr>
<tr><td></td><td></td><td></td><td></td></tr>
</table>

Figure 9.2 Planning assumptions

PLANNING OBJECTIVES	
Department or project ____________ ______________________________ ______________________________ Manager or team ________________	Date ______________ Page____of _______
Short-term objectives	
Long-term objectives	
Expected method	
Expected results	
Change-programme stages and target achievement dates	

Figure 9.3 Planning objectives

SWOT ANALYSIS	
Department or project ____________	Date ____________
____________	Page ___ of ______
Manager or team ____________	
Major change area	
Strengths	Weaknesses
Opportunities	Threats

Figure 9.4 SWOT analysis

No one plan or strategy will provide a 'forever' resolution to the problems associated with change. The process of assessing the need for change and the implications and methods of change, and reviewing the original business and/or company change plans, will be continuous.

RESISTANCE TO CHANGE

Resistance to change is a normal human response. People feel more comfortable with what they know rather than with the expected results of change. The average member of staff will not seek radical change, but will usually cooperate when the reasons for change are clearly explained.

Understandably, the staff will have fears that will initially frustrate changes. These may include fears about:

- a loss of earnings because of a reduction in overtime or bonuses, or changes in shift patterns,
- their skills being made redundant, and thus the potential loss of their employment,
- new technology and the demands that it will bring (this can manifest itself in doubts about their ability to operate the new technology successfully, or to achieve the targeted production norms),
- loss of status (this may occur when departments or companies are merged).

Other reasons for resistance to change that may be seen as trivial by the management but very important to members of staff are:

- a change to the working environment in terms of location or surroundings,
- changes of colleagues.

A manager will occasionally be faced with an employee who resists change in any form whatsoever. Fortunately, this type of attitude is rare, and the great majority of staff will eventually recognize the need for change, even if they initially put up a token resistance. With intransigent members of staff, the company should take care to explain all the relevant factors, including benefits and the imperative for change, and seriously seek cooperation. Failing success, the change should proceed, and the members of staff must either work with the changes or leave the company. Such a course should be a last resort, as it indicates a failure by the management.

Recognized methods of overcoming change are:

- explaining and communicating all relevant facts about plans and the way in which jobs and conditions will be affected,
- explaining the reasons for company policy and the approach to change,
- involving the staff in decision making by seeking their views on change proposals and programmes,
- emphasizing the benefits to be gained from planned change.

MANAGEMENT PERSONAL SKILLS AND DEVELOPMENT

Negotiating

When negotiating change, it is most important that the management avoids, if at all possible, a confrontational situation. In these circumstances, goodwill is lost, and retaliatory moves can result.

Managers who wish to implement change should attempt to negotiate from a position in which both sides have a common interest in progress. This situation can only be achieved through careful preparation before any negotiation meeting.

The consideration of negotiation objectives and supportive arguments is essential, as is assessing the arguments likely to be used by the other side. If a number of people are involved on either side during discussions, it is important that a united front is presented. Any difference of opinion among the management will indicate weakness, and may be exploited.

Negotiations for change can range from informal discussions between interested management and staff to a large-scale formal meeting. In the second case, it is important that accurate records of the meeting are kept.

Negotiations may contain an element of bluff. Either side may propose conditions that are far in excess of those that are expected or required. An example is when a manager states that, in the absence of any agreement, staff will be made redundant, when in fact this is not the case, as alternatives of retraining and redeployment have been prepared.

Obviously, both sides, in making and receiving claims, have to assess the validity of any statements made.

Gottschalk[3] defines negotiation as an occasion when one or more representatives of two or more parties interact in an explicit attempt to reach a jointly acceptable position on one or more issues.

Walton and McKersie[4] state that four types of negotiation exist:

- *Distributive bargaining* exists when the function of negotiation is to resolve conflicts between the parties, one of which has to win.
- *Integrative bargaining* refers to negotiations over issues when no fundamental conflict exists. The situation is therefore capable of a resolution in which both sides will benefit to some degree.
- *Attitudinal structuring* is a process in which one party to the negotiation attempts to influence the attitudes of the other to the benefit of itself.
- *Intra-organizational bargaining* assumes that those directly involved in the negotiation are representatives of their respective sides, and recognizes the important fact that dis-

cussion and debate will occur within both sides in order for a consensus of opinion to be reached.

Decision making

The primary task of a manager in relation to dccision making is to make the right decision at the right time to achieve optimum company performance. A print manager has to make frequent decisions about production technology, methods, staff, customers and finance.

Normally, little attention is given to the way in which these decisions are made. Often, habit or experience will be the resolving factor. Many day-to-day decisions are satisfactorily made in this way. Time alone may prevent any decision being made other than one based on experience. However, a different approach is required when change is managed. Considerable attention should be given to the problems arising from change, as no change implementation can be successful without the problems being resolved first. When there are various options for the resolution of a problem, Baye's rule often operates, which states that when it is necessary to decide between options, the one with the greatest chance of success should be chosen.

Decision making is an essential skill when it is applied to crises in production, human resources, and the wider aims of improving the company's performance.

Success in decision making will be greatly enhanced if the manager makes use of the abilities and knowledge of each department.

Management self-development

Management responsibility for training does not stop with the workforce. Managers need to consider self-development as part of their progress. All managers have a responsibility to develop themselves continually, and to keep up to date with knowledge and practice. Often, it may be necessary to acquire new skills. Print managers should concentrate on development in the important skills of:

- performance appraisal,
- analysing human-resources provisions,
- understanding and reviewing company objectives and their demands upon management,
- planning and developing strategies to accommodate change.

Projects and case studies can be a valuable form of training, especially when analysis and detailed reports are required. Other methods are attendance at lectures, seminars and short courses, memberships of management institutions, and selected reading.

Management responsibility for staff training

The training of production staff and other staff is a prime responsibility of management, and investment in specific training is essential for the stability and progress of a company. Some companies see no need for training as they 'only employ skilled people'. While, at first sight, this approach may appear to be satisfactory, it may leave the company in a position in which it is unable to respond quickly to the demands for new knowledge and skills created by rapid change.

Training for training's sake is of little use to a company. A careful analysis of immediate and medium-term requirements to meet change is essential. The analysis of training needs should include the following factors:

- any expected market changes that may require the acquisition of new skills,
- the human-resources strengths of competitors,
- developments in technology,
- planned financial resources,
- legislation that is likely to affect future plans,

- skills training whether it is required to update existing skills, or to cater for an entirely new field,
- training, whether it is required to overcome inferior working methods or interpersonal factors.

Time management

Efficient time management is a facet of personal organization. Time, or the lack of it, can be one of the biggest problems facing the print manager, as lost time can rarely be recovered. The majority of managers take time for granted. However, almost every manager could make better use of his/her time. The key to the efficient use of time is organization. Work should be allocated to its correct category, e.g. action, file, pending or priority action. An untidy desk, on which the manager is constantly searching for paperwork, is a certain way to waste time.

All managers can gain valuable information about their use of time by keeping a diary of individual items of work and the time occupied on each. Analysis will indicate any jobs that could be delegated elsewhere, and suggest time prioritization for the remainder.

The prioritization of work is very important, and any tendency to deal with the easiest matter first should be resisted. Delegation is probably one of the most important and effective ways of saving time, and it is a vital skill for all managers.

Delegation

Delegation is a skill that enables a manager to achieve aims and objectives through people. No manager has the time, and probably skill, to carry out all the management tasks personally. Thus the ability to delegate authority becomes very important. Delegation is much more than just giving work to other members of staff. It should mean assigning authority and accountability for all the decisions taken within the authority delegated, although the manager always retains overall responsibility.

Advantages of delegation include the following:

- Correctly applied delegation can improve the efficiency of a department or company through staff participation and motivation and improved job satisfaction.
- Total reliance on one or two individuals is reduced, thus resolving the problems of continuity in cases of holiday or absence.

An essential ingredient of delegation is management trust. It is pointless for managers to delegate work and then insist that it is carried out in exactly the same manner as the manager would have used him/herself. Management trust can produce positive results, but excess control results in employee uncertainty, apathy, and possibly resentment.

Meetings

Correctly controlled meetings are an effective method of communication, as verbal discussion can save a large amount of paperwork and time. However, large meetings often achieve little, and can be expensive. Thus any manager arranging a meeting needs to carefully consider:

- what is to be discussed during the meeting,
- the purpose of the meeting, i.e. what is to be achieved by the meeting (is its purpose to decide action, inform or persuade?),
- any formation of teams or subcommittees.

Attendance at meetings should be confined to those who have something positive to offer, and who are in a position to influence results. The chairperson should maintain control to ensure that only the published business is discussed. Any plans, documents or other data requiring study should be distributed in time for it to be read before the meeting. Time spent in preparation for meetings will prove an investment, as the success of the meeting will be in direct proportion to the level of preparation and the control exercised. All meetings should last for an allotted time, the time being just sufficient to allow essential business to be covered. This will have the effect of diverting attendees away from small talk or trivia that can be dealt with elsewhere.

CUSTOMER CARE

The main aim of customer care should be to develop customer relations to the point of mutual trust and respect. Good customer relations are not easily achieved, and they take time and effort to develop. However, they are quickly damaged in the event of problems arising from indifference or inability. Major strengths for any printing company are a loyal base of customers, and a reputation for providing quality in all aspects of its work.

Customers may find it difficult to differentiate between printers' qualities for similar types of work. In this situation, a customer may be influenced by company attitude and approach.

Key points of customer care are:

- impressing on the customer that his/her work is important and valued,
- communicating efficiently through company representatives,
- helping the customer to obtain what he/she needs,
- providing adequate management support for marketing activities,
- providing specific training in customer care for all the employees who are likely to meet customers,
- handling all enquiries smoothly,
- handling complaints efficiently and sympathetically,

- combining technological knowledge with courteous service,
- talking to customers in a way that they understand,
- handling accounts diplomatically,
- making occasional customer calls, not to sell, but to discuss printed products and any problems of performance etc.

COMMUNICATIONS

Verbal presentation

Verbal presentation has a disadvantage, in that it consists of one-way communication. It is therefore only suitable for situations in which information is given, as it allows little discussion or questioning. It is often associated with an authoritarian approach. Verbal presentation can be assisted by the use of an overhead projector. This is one of the most popular visual aids, as it allows images to be projected, free of distortion, in normal lighting conditions. Its use allows all those present to see the key issues displayed clearly. Line charts, bar charts, Gantt charts and PERT charts are suitable for presenting data. Colour coding by overlay can enhance the impact of the presentation. Particular care should be taken when using any form of projection to ensure that no part of the projection is obstructed. The value of the information will be lost if people cannot see it. Computer printouts can be used to present financial data, project plans, statistics and records. Bar charts, histograms and other charts can add depth to the presentation.

Seminars

A seminar in which a group of people discuss various issues is an effective means of two-way communication. Key points for discussion are determined, and all present have the opportunity to participate. The manager should take a leading role in controlling the discussion, to ensure that all the participants feel free to contribute, and that the discussion is not dominated by any one individual. The success of a seminar will lie in the pooling of knowledge and views. Also, participants develop a better understanding of the problems involved. It is important to identify objectives and the limits of discussion clearly, and to ensure that any required information is freely available.

Reports

A carefully constructed report is an excellent way of introducing and communicating issues that require considerable thought and response. A report should be a working document that is easy for all its readers to understand. The appearance of the report and its arrangement are important:

- The report should be designed so that it has a neat and clear appearance.
- It should have a title that clearly indicates the purpose of the report. The date and the name of originator should be included.
- Information about the contents by chapter or item number, should be included.
- An introduction should be included that gives the reasons for writing the report, states its aims, and indicates any response required.
- The contents must be structured in logical sequence.
- A summary and conclusions must be included. This is most important. Although the readers will wish to see the writer's method of arriving at the conclusions, the conclusions will almost certainly be the subject of discussion and response.

The following are important points to consider when any type of presentation proposing change is prepared:

- exactly what is to be presented,
- the key evidence to support the case or proposals,
- the reasons why change must occur,
- how it will affect those involved,
- the benefits of change to those involved,
- the possibility of alterations in response to other proposals,
- any arguments against the proposals.

EUROPE

The European Community did not have a good start. The 1990s were planned to be a period of integration and union. Unfortunately, visions of fast progression in the 1990s have not been fully realized, although it is assumed that the European Community's prime purpose is the achievement of economic integration. Article 34 of the charter[5], covering nontariff barriers to trade, states the following:

> Quantitative restrictions on exports and all measures having equivalent effect shall be prohibited between member states. Member states shall, by the end of the first stage at the latest, abolish all quantitative restrictions on exports and any measures having equivalent effect which are in existence when the Treaty enters into force.

However, barriers still remain, and they have been difficult to overcome. Examples are:

- state subsidy to industry,
- differing technical and product standards,
- differing advertising rules,
- product licensing and price regulation,
- purchase restrictions that favour national suppliers.

However, it is not suggested that a negative attitude should be adopted to European trade. Trade barriers are beginning to ease, and will possibly disappear within the medium-term future. The key requirements for successful entry into European trade are fast becoming service, quality and price. These issues are obviously those facing printers in the UK, but, with the attention being paid to these issues by European printers, a threatening situation could develop for some UK printers.

There appear to be no simple ways of entering European trade, but it does seem that employing local staff, who obviously have an in-depth knowledge of the market situation in any of the locations in question, is a distinct advantage. Partnerships with European companies can also be considered.

Irrespective of the circumstances, it will be wise to have a local office in Europe. Although distance communications are effective, efficient customer care becomes more difficult over long distances. The alternative of frequent commuting can be very expensive and restrictive in terms of time.

The widespread use of the English language has, to a large extent, removed the difficulty arising from language differences. Even so, fluency in another language will assist two-way communications and resolve any perceived integration problems. Obviously, the European market provides considerable opportunities for some UK printers. Any threats can probably be overcome by a careful approach to customer care. Any printing company contemplating entry into the Europe marketplace must review its organization and working methods. The technology in use in the UK is no different, usually, from that used in Europe. However, methods of use in Europe are often different, with considerable attention being paid to organization and productivity. Overcapacity is avoided, and efficiency ratios of 90% are common.

Printing companies considering entry into the European market should first evaluate price structures, market opportunities, distribution costs, communications and the transfer of capital or profits.

The great majority of UK companies see no need to attempt entry into European markets, but, as integration progresses, it may become essential if they are to survive. Even companies with no immediate interest should keep up to date with knowledge about methods of operation and any significant developments.

An increasing cooperation between France and Germany is apparent. Both of these countries currently offer lower prices for a wide range of print work. This has particular significance for UK printers, indicating that work may move to Europe in the absence of price adjustment.

European legislation includes a Charter of Rights of Workers[6], which stipulates:

- freedom of movement for all workers of member countries,
- a right to adequate health and safety protection,
- adequate social protection.

This in theory, means that print workers are free to work in any member country. This presents new challenges for print managers. In the foreseeable future, a new type of manager will be required. He/she will need to be able to operate throughout Europe, be fluent in two or more languages, and understand cultural differences.

SUMMARY

- Change within the printing industry is inevitable, and it must be faced.
- Radical change presents the manager with problems that require particular skills to resolve.
- Efficient management of change is essential for survival and progress.
- The development of the correct management style is crucial.
- Managers should practice self-evaluation and development.
- Staff involvement through consultation will help with any changes planned.
- The accurate assessment of the need for change is very important.

- Plans and strategies for change must be carefully thought out.
- An evaluation of all the implications for all departments and staff is required.
- When it becomes necessary to reduce staff levels, a sympathetic approach should be adopted.
- Strategies for change should follow a carefully planned sequence.
- Change can be used to increase job satisfaction.
- Resistance to change is a normal human response. However, it may be resolved by clear and truthful communication.
- A manager should pay attention to the skills of negotiation, decision making and presentation.
- Management has a particular responsibility for staff training to accommodate change.
- The correct delegation of duties is essential to achieve objectives.
- All managers must develop the ability to prepare and control meetings carefully.
- Customer care is now more important than ever before.
- Careful time management to allow the evaluation of company performance is essential.
- Companies considering entry into European markets should carefully evaluate all the issues involved prior to planning.

ACTIVITIES

- *Activity 9.1*: Prepare a change plan for one of the case studies in Chapter 10.
- *Activity 9.2*: Outline the key abilities for any manager wishing to develop and manage change.
- *Activity 9.3*: Prepare a self-development plan for a manager.
- *Activity 9.4*: Identify any management-of-change issue that is not dealt with in this chapter.

REFERENCES

1 **Birkenshaw, J W and McArthur, R (Eds.)** *Ten Year Technology Forecast for Printing and Publishing 1991–2001* Pira International, UK (1991)

2 **Roethlisberger, E and Dickson, W** *Management and the Worker* Harvard University Press (1939)

3 **Gottschalk, K** 'The background to the negotiation processes' *in* **Torrington, D (Ed.)** *Code of Personnel Administration* Gower, UK (1973)

4 **Walton, C and McKersie, J** *Pay Productivity and Collective Bargaining* Macmillan (1973)

5 *Article 34. European Commission's White Paper on 'Completing the Internal Market' 1992 – supported by the EC Court of Justice* Command Paper 1934, HMSO

6 *Community Charter of Fundamental Social Rights of Workers 1989* Command Paper 1934, HMSO

- *Administration*: There are seven staff. Two staff are engaged in estimating, sales and invoicing. Both are slightly over 30. They have some sales experience, but little technical knowledge.
- *Work reception, general reception, and some customer liaison*: There are two staff. One is female, and acts as the general receptionist, and passes work received to the appropriate area. This is usually origination. The other writes works orders etc.
- *Finance*: There are two staff, both of whom have some experience in manual financial administration. One is 35 and works on salaries, accounting etc. The other acts as the company secretary (the company secretary also acts as the office manager). He is an experienced production manager, but his main experience is in the press area. His business experience has been gained by self-development since he became a manager about seven years ago. He is 43.
- *Origination:* There are six staff. One is skilled in most origination processes, and is 31. Three are mainly self-taught, but have had some company training. Their ages are 34, 41 and 22. There are two trainees, both under 18. The equipment in the origination area includes one recently installed A2 camera, and one very large camera that is about ten years old. There is a range of studio equipment that includes four light tables, processing equipment and hand printdown frames. There is some autoprocessing equipment. However, most plate preparation is done by hand, as the autoprocessing seems to be unreliable. All the staff except the trainees follow the work through all the stages, from the receipt of artwork to planning, camera, film assembly and printdown. This arrangement is designed to increase employee job satisfaction and customer care.
- *Printing*: There are six staff. Two are employed on a double day shift on a 2-colour press. Each shift is approximately 9h long. Both staff are experienced and qualified, and they are 29 and 34. Two staff are employed on single-shift working (a shift lasts 9h). One is experienced, but mainly self-

taught, and the other is a trainee who is 22. Two general assistants are employed in the press room. There is one on each shift. Both are over 44. The technology includes one RA2 2-colour press (a Heidelberg, three years old), one RA2 single-colour press (a Heidelberg seven years old), and one RA3 single-colour press (a Ryobi, five years old).

- *Finishing*: There are six staff. One operates an SRA2 combination folder that is four years old, and has considerable experience of not only folding, but also the majority of finishing operations. This person is 41. There is one operator of a guillotine that is 42 in programmed and two years old. This person is experienced, but mainly in cutting. Two staff are employed on a Müller stitching and trimming line. One is 34, and is experienced, and is training the other, who is 21. One member of staff, aged 60, is employed on packing and general duties. One is employed as a general assistant, and is qualified to operate the company's fork-lift truck. He is within two years of retirement.

The company realizes that some technology may not be suitable for the work being produced, but is convinced that the problems that the company is experiencing are not related to technology alone. Two special meetings to discuss the situation have failed to produce an agreement on the correct strategy to resolve the situation. The existing situation has led to a decline in the company's financial position. The company is consequently reluctant to invest in new technology until an improvement in productivity and profit has been achieved.

The company has now called on an external print consultant to advise them. What are your opinions?

CASE STUDY 2

This study concerns a large public organization that places a considerable quantity of print orders to a value of approximately £1.7m/annum. All printing requirements are placed externally, as the organization has taken a policy decision not to invest in an in-house printing unit. The main reason for this decision is the very large financial investment that would be required to carry out all the work satisfactorily.

The work is handled by placing print requirements with one of five design agencies, which then prepares the artwork etc. Depending on the print specification, the agency then contracts a printer to complete the order. For some work, the design agency arranges for origination, and, for other work, the origination is carried out by the printing company. The public organization's two print buyers' knowledge of printing production and processes is sufficient to enable reasonably technical conversation to be held with the design agencies. However, the volume of work required is increasing, and there is a change in the work type, which now requires creative design and work of high quality. The change in work has, to some extent, been accommodated by the design agencies. Three of them have been able to satisfy the requirements to a greater degree than the other two. However, quality and prices are giving cause for concern. Invoiced prices tend to exceed estimated prices, by considerable margins in some cases, and delivery dates are not always met. Currently, the organization has completed an order for a large prestige production, for which the quality of work has been criticized, and the organization is therefore unhappy to see the work published and distributed. There is a general dissatisfaction with a number of items, such as illustrations, the reproduction of typefaces, and text reversals.

Although this is not the only job that is causing concern, it illustrates a growing trend in that, it is one that has been very costly to produce. This has resulted in a meeting within the organization to decide what to do about the situation, and to assess possible reasons for the trend prior to consultation with various design agencies. What is certain is that the present circumstances cannot continue.

What are your views about the issues that should be discussed at the meeting, and the steps that should be taken to improve the situation?

CASE STUDY 3

A UK printer entering a new field of production has acquired a contract for work that requires print that is numbered and perforated. The printer has no existing facilities for these operations, but price levels mean that they must be achieved in line.

The printer has purchased a second-hand press on the understanding that numbering and perforation is easy to set up. This is most important, as the press will not be used only on the contract work.

The supplier of the press has successfully installed other presses with the same specification in the past.

The technology has been purchased after discussions with a sales representative of the supplier, and the perusal of literature that is specific to the item of technology provided by the supplier.

Subsequent to the installation and commission of the technology, the printing company has come to the conclusion, after many production problems, that the technology is incapable of operating as described in the literature and described verbally by sales staff. The sales staff said that the operation of the technology was straightforward, and that little time would be required to set up and make ready, and, further, that only average experience would be needed by staff to achieve successful operation.

Following continuing problems and further discussion, the printing company has refused to accept the technology, and has paid the 50% of the purchase price that was due on completion of installation and commission. The supplier has tried to overcome the problem by providing staff training, however this has not been successful. This may have been due, to some extent, to staff attitudes. They have lost confidence in the technology.

The printing company has demanded that the technology be removed from its premises. The supplier has refused, and has sued for full payment, with costs and interest.

The case is listed to be heard in the Crown Court in the near future.

The printing company (the defendant) has entered a counterclaim for lost production and damage to business opportunities.

What are your views of the management of this situation, and how do you think it should have been handled?

CASE STUDY 4

A newspaper printer and publisher with a turnover of £3.2m/annum producing morning and evening papers six days a week has experienced a radical change in production demand.

The production method for one of the newspapers has changed from rotary-letterpress to web-offset. In the initial stages, the web-offset printing was confined to black with spot colour. However, the advertising business has begun to contract, the clients require full colour in preference to spot colour, and quality demands continue to increase.

For the paper to survive, it has become obvious that financial investment in new and advanced technology is required. This has been agreed by the company's board, and it has been achieved through strategic planning. It has also been recognized that the changes in technology require a review of human-resource provision if efficient production is to be achieved and maintained. The initial stage in resolving this situation entailed 'head hunting' for a suitably qualified technical production manager. A manager from another newspaper producing full colour has been employed.

Unfortunately, this has been insufficient to resolve the situation in itself, as the length of the production process often exceeds 20 h. Further analysis indicates the need for additional training for all the staff involved in press production.

The staff details are as follows:

- *Day working*: The staff consists of a technical production manager, three reel hands, and six printers. The reel hands and printers are supervised by a production overseer. The production overseer has long experience in rotary-letterpress printing, and approximately four years' experience in web-offset printing in black and spot colour. Some multicolour experience has been gained through recent training by the new technical production manager. The three reel hands only have knowledge of paper handling. However, two of them are interested in development if training becomes available. The other is quite happy as he is, and does not seek any change. The production overseer is 35, and the two reel hands ready for development are 50 and 48 (the one who does not wish to change is 50). The printers' ages range from 24 to 37, and they all appear to be keen to develop their skills.
- *Night working*: The night shift consists of a production overseer, an assistant, six reel hands, and eight printers. Extra manning at night reflects the increased production demand during this period. The night production overseer has a little more knowledge of multicolour web-offset printing than the day overseer, but admits that further knowledge is required. His assistant is less qualified, but is keen to advance. All the reel hands appear to be content with their situation, but would develop if necessary. The printers have been retrained from rotary-letterpress printing, and from spot-colour web-offset printing. Three of these printers are difficult to handle, because of their industrial attitude. The remainder wish to develop, but not without wage negotiation. The night overseer is 37, and his assistant is 29. The printers' ages range from 21 to 36. The reel hands, except for one, are all over 48. All the reel hands have been employed for 12–14 years, two of the printers for five years, and the remainder for less than two years.

Under the current circumstances, considerable investment has been made in new technology, and a new technical production manager has been employed, but unfortunately the problem of loss of quality and production downtime has not been completely resolved. There is a human-resources situation that requires urgent action. What are your views on this?

CASE STUDY 5

M Printers have traded for many years as small offset printers producing good-quality stationery and similar work. When larger, and often multicolour work is offered, the work is subcontracted to other printers, depending on the work type and production demand. Subcontracting work is not regarded as satisfactory, but as preferable to refusing the work with the possible consequent negative customer perception.

Two issues are now beginning to cause serious concern for M Printers. First, the demand for multicolour work that is beyond their capacity is increasing, and, second, customer complaints about the quality of work have reached an unacceptable level.

An analysis of the complaints reveals that they mainly relate to the work that has been subcontracted, over which M Printers have little control. Consequently, administrative and other work is affecting profit margins. M Printers have now decided to investigate the possibility of producing all the work themselves, but they are not clear about the implications of such a move. The company has therefore decided to use the services of a consultant to assess the feasibility of accommodating all the work within the company.

In your view, what information will the consultant require to assess the feasibility of the plan, and present M Printers with the options and their implications?

CASE STUDY 6

A large printing organization, X Printers, has appointed one of its managers to organize a printing company that it has taken over. X Printers' directorate consider that the company has potential, and it has therefore decided to allow one year for improvement.

The acquired company is involved in short-run magazine and trade printing. However, a consistent trading loss has resulted in the takeover. Although the acquired company has sufficient work to providc a profit, it appears that, following the retirement of the managing director 18 months ago, the financial situation rapidly declined to a point at which liquidation was a real possibility.

At first sight, it appears that uncoordinated administration and department management is one of the key problems.

The company has three 2-colour presses. One SRA1 is four years old, and the other two SRA2 presses are two years and six years old. They have all been purchased outright, and there are no loans or other financial liabilities. There is also one SRA1 4-colour press that is 12 years old, and one SRA2 single-colour press that is three years old.

The finishing facilities include one 42 in and one 32 in programmed guillotines. Both are over seven years old, but are in good condition. There is an automatic, 6-station stitching and trimming line that is four years old. There are two combination folding machines, one of which is SRA1 and is 1.5 years old, and the other of which is SRA2 and six years old.

All the origination is purchased externally from a repro house close to the company. Facilities are available for plate processing, but they are rarely used. The premises are old, but are in good condition. Printing takes place on the ground floor, finishing on the second floor, and administration on the third floor.

The staff situation is as follows:

- *Administration*: There was a manager, whose employment has been terminated by agreement. The deputy manager, aged 46, has considerable technical expertise, but no formal business training. There are a company secretary, three other staff who cover all the financial and other administration, and one receptionist/typist. The company secretary is 48, and has been with the company for ten years. The others are all under 30, and have less than two years' service with the company.
- *Pressroom*: The pressroom employs 14 staff, seven on each shift. Two shifts operate, each for 8 h. All the printing staff are qualified through formal training. Four staff are over 52, with company service of 14–18 years. Six are over 40. Four have been with the company for approximately five years, and the remainder for less than two years.
- *Finishing*: There are ten finishing staff. Only four have formal qualifications, but they all appear to operate efficiently within their areas of responsibility. Five of the finishing staff are over 52, and have 8–10 years service with the company. Three are aged between 25 and 37, and have an average of four years' service. The remaining two are under 20 and they have less than two years' service with the company.

Obviously, the manager appointed to deal with the situation faces some difficult problems.

What are your views of the problems, and the potential methods for resolving them?

CASE STUDY 7

This study relates to the in-house printing unit of a large mechanical-engineering organization. The work is composed of all the print requirements that are needed for the operation of a large international organization, except multicolour production. Currently, all the multicolour printing is placed with external printers. The decision to print externally is usually based on the opinion of the origination section.

However, the demand for multicolour work has increased from 25% to 55% of overall production. Apart from the manager and the estimating and costing clerks, all the staff are flexible, and in fact work in various production areas, i.e. origination, printing and finishing, on a change basis. Each change lasts for one month. The objective of this arrangement is to create job interest.

Because of a varying demand for work at short notice, a permanent overtime scheme is in operation, in which all production staff are paid overtime each week on the understanding that overtime must be worked when requested.

The printing unit employs 26. Four are under 21, one is 62, three are in their mid fifties, six are over 40, and the remainder are 20–40. Two costing clerks and one estimator handle all the financial and administrative matters required by the organization. These staff are increasingly concerned about the imbalance between estimated and actual costs. The older staff appear to have more business awareness than the younger staff.

The manager considers that production is below that which one would normally expect, and he is very concerned that the unit appears to be unable to produce multicolour work. Unfortunately, he is not quite sure how to establish the facts and prepare a strategy to deal with the situation, as many different reasons for the situation are given by the production staff.

The organization is willing to invest in larger and more advanced technology, but it has serious doubts about the skills of the production staff, and fears that the introduction of new technology will cause further problems. Another consideration is the large expenditure on work passed to other companies. Naturally, these companies are well known to the origination section.

All the production staff have received formal training. The relations between the management and the production staff are only adequate.

The in-house printing unit is presently subsidized, as it is regarded as a service. However the organization is now insisting that the unit must become cost-effective within one year, although no real profit is required.

What are your views about this situation, and the methods that should be used to overcome the problems?

CASE STUDY 8

Z Printers are to cease trading in the near future, and this provides the opportunity for B Printers to take over contracts for some of Z Printers' work. B Printers are already working for the publishers of the work, who are willing to place the contracts with B Printers, provided that assurances of efficient production and quality standards are maintained.

The work is as follows:

- *Publication 1*: This is a weekly specialist publication (32 pp text, A4, 4 pp covers printed 4-colour process, remainder of work (text) black only, line and tone). The paper is 100 g/m^2 blade coated. The finishing is 2-wire saddle-stiched. The first 9000 copies are bulk packed for delivery to a distribution company. 13 000 copies are despatched to subscribers. The work is required each Friday, and it consists of 22 000 copies of the publication.
- *Publication 2*: This is a newssheet, published once a month, (12 pp A3, printed 2-colour throughout, on 100 g/m^2 white cartridge paper). The print run is 3250 copies, and it is delivered in bulk to the publishers.

  The human resources and technology available at B Printers are as follows:
- *Administration*: This consists of a managing director, a secretary, two costing and estimating clerks, and an office junior.
- *Studio/origination*: This consists of four staff, who are all qualified and experienced. There are two A2 vertical cam-

eras, four light/assembly tables, and three printing-down frames.

- *Pressroom*: This has five staff, including a general assistant. All the printers are experienced and qualified. The equipment consists of a Heidelberg MOZP 4-colour press, a Heidelberg MO 2-colour press, a Heidelberg GTO single-colour press, an R30/95 single-colour press and a Heidelberg D/Crown Letterpress press (used occasionally for creasing, embossing etc.).
- *Finishing*: There are six finishing staff, four of whom are experienced and qualified. The others were trained within the company, but are efficient workers. All the staff are flexible (in terms of working areas). The equipment consists of a Perfecta guillotine (programmed 32 in cut). There are two SRA1 Stahl combination folders, and a Linotype Trimstitch with five stations. There are fork-lift-truck, packing and baling facilities.

The company management are due to meet and discuss the work opportunities and the implications arising. Currently, the company is profitable, but it has little finance to invest in new technology. Initial analysis indicates that the company is operating at approximately 71% of its capacity. This alone is causing some concern.

Room for expansion exists if necessary. However, some building work would be required, as the expansion area is inconveniently situated without easy access. Currently, all the paper is stored in the pressroom.

What are your views about the management action required to reach the correct decisions?

CASE STUDY 9

This study relates to a production dispute between a printer and a customer.

The customer has directly placed an order with the printer for a 48 pp + 4 pp-cover booklet. The cover is printed 4-colour process, and the text in two colours throughout. The work is 2-wire saddle-stitched. The paper is 115 g/m^2 art paper.

The dispute is about the cover, which is printed on 240 g/m^2 white cover paper. The customer supplied line artwork, and three transparencies for the cover. The specification included three PMS colours. However, the customer also specified 'print by 4-colour process'. Because of extreme urgency and price considerations, the customer did not require proofs of the cover, although ozalids of the text were submitted.

Following completion of the work, the customer at first refused to accept it, claiming that the specified PMS colours had not been matched correctly. The printer pointed out that the PMS colours in question were difficult to match precisely using 4-colour process printing, and that they should have been printed as self-colours, although this would have increased the cost of the job. The customer still maintained that the colours requested could be obtained by 4-colour process printing, but agreed to accept the work on condition that a substantial reduction in price was made. After prolonged discussions, the printer accepted that two options existed: first, to enter litigation to recover the money owed, and, second, to accept that a loss was inevitable, and settle for a substantial reduction in the invoiced price.

What are your views of this situation, and how can a printing company avoid these situations, especially when profit margins are small?

CASE STUDY 10

A large construction organization with depots in various parts of the world has identified a need to review some of its services. One of these services is a large in-house printing unit. The unit operates independently, but it is not required to make a profit. The in-house unit offers a complete printing service to all the depots, and the range of work is very wide.

During the course of review, it has become obvious that many of the depots would prefer to have their print requirements produced locally, rather than centrally through the main office. This situation has led to considerable discussion by senior management, who have differing opinions as to the value of the in-house unit, and whether it should be retained, or whether all work should be placed externally. Price checks have indicated that some work could be produced at less cost externally. However, administration costs are likely to increase, and any reduction in work in the in-house printing unit will have serious implications for the unit's viability. Some data relating to the unit's operations is as follows:

- A crosssection of orders taken at random were as follows: continuous stationery (seven orders), general stationery (19 orders), information services (9 orders), multi-page-colour (11 orders).
- The annual turnover in terms of classes of work is as follows: multipage: 38%, continuous stationery: 13%, general stationery: 22%, information services: 27%.
- The equipment includes Apple Mac stations linked to a flatbed scanner, advanced photocopying systems, manual film-assembly and printing-down facilities, eight printing presses, ranging from small offset to A2 multicolour, various folding, stitching and trimming facilities, including a stitching and trimming line. Administration is fully computerized, as is the delivery-recording system. The resale value of all the technology is estimated to be £1.5m.
- 18 production workers are employed. All except one work full-time. The age average of 43 is rather high. However, many workers have long service with the unit, and considerable staff flexibility exists within specific production areas. The unit has, in the past, received adequate finance for investment in new technology. However, this issue is now

under review. Six administration staff are employed; this includes the general manager.

- Financial management is covered within the unit (although it reports to, and is responsible to, the main organization), and it appears to be of a high standard.

The organization has to reach a decision about the future of the in-house printing unit, but, before any decision is taken, a range of information about the unit's operation is required. What are your views about what information is required, and how it is best obtained?

CASE STUDY 1: ACTIONS

- Determine the financial position.
- Identify the existing management structure, style and method of operation.
- Establish whether financial problems arise from production, administration, or both.
- Interview administration and production staff to develop staff profiles, skills and attitudes.
- Establish whether there are any links between wage and salary increases and the overtime worked.
- Analyse overtime in relation to downtime.
- Establish the causes of downtime by reviewing work methods, and analysing production records.
- Review the effects of flexitime on production, and, in particular, departmental approaches to flexitime. Evaluate the effects of flexitime on the late delivery of work.
- Identify the main reasons for the late delivery of work.
- Assess technology and other equipment in terms of its, suitability, age and condition. Identify any changes required, and their financial implications.
- Prepare the conclusions of the study, and the options available and their financial implications.

CASE STUDY 2: ACTIONS

- Assess the knowledge of the print-buying staff, through consultation, to establish whether any training is required.
- Establish the method of design contact and liaison, including verbal and written technical specifications.
- Identify the main areas and subjects of complaints, to see whether any particular problem is obvious.
- Examine the relationships of the design agencies with the printers and the repro houses, and identify the method or criteria by which they are selected.
- Determine the level of control by the design agency over print production. Assess the impact of control on the nature of the complaints.
- Examine the methods of working to determine the number of corrections, and at which stage of production they are made.
- Draw conclusions, and decide on and implement action.

CASE STUDY 3: ACTIONS

The outcome was that the case did not reach court. The supplier agreed to, at his expense, replace the numbering and perforating attachment to the press, as it was later found to be faulty. The cost of the press was reduced.

Training was arranged for the staff who were to operate the equipment. This overcame their lack of confidence arising from the first frustrated attempts at production.

Considerable trouble could have been saved if:

- the supplier had tested the press and equipment before sale and installation,
- the printing company had insisted on a production demonstration before purchase,

- the staff who were to operate the equipment had been informed of the importance of the installation, and training had been arranged before installation.

CASE STUDY 4: ACTIONS

- The board should hold discussions with the technical managers to establish their positions and views.
- Identify areas of expertise or skills that are required to meet production demands.
- Observe both shifts in operation to assess the production situation.
- Interview individual staff from both shifts to prepare a skills and attitudes profile.
- Identify any skills deficiency in each shift, and evaluate the possible benefit of transferring staff from one shift to another.
- Identify the need for any staff change or addition.
- Prepare an in-house or external training programme to cater for the identified needs.

CASE STUDY 5: ACTIONS

- Assess the value of the subcontracted work, and its contribution to turnover and profit.
- Identify the main reasons for complaints, and any connection with individual printers or types of work.
- Evaluate the existing capacity, and that required to cover increased demand.
- Review company procedures for subcontracting work, with respect to technical specifications etc.
- Consult the staff.
- Identify what technology and human resources are required to meet any increased production, and the financial investment required.

- Identify any training required.
- Prepare and discuss the options for the company. Decide on the correct actions, and implement them.

CASE STUDY 6: ACTIONS

The financial position has already been established, and it is apparently the main cause of the current situation.

- Review the financial administration system and its operation, initiating adjustments or replacements where necessary.
- Review the management structure.
- Consult the staff.
- Prepare staff profiles to identify skills and attitudes, and any training needs.
- Evaluate the condition of the technology, and its suitability for existing and potential work.
- Study work methods and times, including work instructions and supervision.
- Review the system for recording production times, and identify the reasons for downtime.
- Prepare an action plan that covers the following:
    - the management structure,
    - administration,
    - human resources,
    - the technology,
    - production,
    - performance monitoring.

CASE STUDY 7: ACTIONS

- Review the management structure and the methods of operation.

- Consult the staff.
- Identify the suitability of the technology for existing and potential work.
- Establish the effects of staff rotation on production.
- Identify any need for new technology, and its financial implications.
- Analyse the overtime worked in relation to that paid for.
- Study the production methods and techniques.
- Identify any training requirements.
- Analyse the results, and identify the options available. Decide on the correct actions, implement them, and monitor the results.

CASE STUDY 8: ACTIONS

- Establish the financial facts relating to the new work.
- Identify the volume of the new work in relation to the existing capacity.
- Identify the implications of:
    - the financial investment required,
    - the required investment in technology,
    - the human resources,
    - the working space and the environment,
    - the purchasing of materials and their storage.
- Analyse the results, make the decisions, and implement them.

CASE STUDY 9: ACTIONS

This study relates to a situation that one would hope would never happen. Unfortunately, these situations do arise, and they often have their origin either in genuine customer ignorance about the limits of printing processes, or in insufficient customer care by the printer. Correct printer and customer liaison would have identified the fact that, while the majority of PMS colours can be satisfactorily reproduced by the 4-colour printing process, some colours can prove difficult, and would, for quality reasons, be best produced as self-colours. The customer would then have had a clear choice of accepting the nearest match achievable by 4-colour process printing, or paying for additional work.

Obviously, in this case, the absence of proofs played an important part in the problem, and it in fact added to costs, instead of reducing them. An important point to consider is that, when there is an error, the further production progresses, the more expensive the error becomes to remedy.

CASE STUDY 10: ACTIONS

- Analyse the type and volume of work by individual depots.
- Identify the reasons for individual depots wanting to print locally.
- Establish prices for local print, and compare them with in-house-unit prices.
- Assess whether any financial savings are possible or probable.
- Analyse the reasons for depots wanting to print locally to see whether their needs can be satisfied in-house.
- Very carefully consider the effects on the unit and its staff of any decision to terminate all or part of its operations.
- Identify all the advantages and disadvantages of retaining, disposing of, or reducing the size of the unit.

Recommended reading

Chapter 1 Printing-company organization: the role and scope of the printing manager

Drucker, P *The Practice of Management* Longmans, UK (1968)
Humble, J *Management by Objectives* Gower Press, UK (1980)
Drucker, P *Management – Tasks, Responsibilities, Practice* Butterworth – Heinemann Ltd, UK (1992)
Drucker, P *Management for Results* Pan Books, UK (1967)
Urwick, L *The Golden Book of Management* Newman Neame, UK (1956)
Barnard, C *The Functions of the Executive* Harvard University Press, UK (1938)

Chapter 2 Print marketing and sales

Bates, M *Marketing* Macmillan, UK (1990)
Cannon, T 'How to win profitable business' *Sunday Telegraph* UK (1987)
Foster, D *Mastering Marketing* Macmillan, UK (1992)

Chapter 3 Financial aspects of print management

Watts, A *Business and Financial Management* M & E – Pitman, UK (1987)
Harper, A *Management Accounting* M & E – Pitman, UK (1991)
Hingley, W and Osborn, F *Financial Management Made Simple* Heinemann, UK (1990)
Secret, M *Mastering Spreadsheets, Budgets, and Forecasts* Institute of Management, UK (1990)

Chadwick, L *Financial Accounting* Prentice Hall, UK (1985)
Anon *Estimating* British Printing Industries Federation, UK
Anon *Costing* British Printing Industries Federation, UK

Chapter 4 Human resources

Walker, J *Human Resource Planning* McGraw Hill, UK (1980)
Vroom, V *Work and Motivation* Wiley, UK (1976)
Thomason, G *A Textbook of Personnel Management* Institute of Personnel Management, UK (1991)
Dale, B and Cooper, C *TQM and Human Resources* Blackwell, UK (1993)
Davey, D and Harris, M *Judging People* McGraw Hill, UK (1988)
Bowey, A *A Guide to Manpower Planning* Macmillan, UK (1974)
Holmes, K *Pira Guide to Total Quality Management* Pira International, UK (1992)

Chapter 5 Printing technology

Babbage, C *On the Economy of Machinery and Manufacture* Knight, UK (1985)
Burness, B *New Technology in Context* Gower, UK (1988)
Jurah, J *Quality Control Handbook* McGraw Hill, UK (1988)

Chapter 6 Print production and operations management

Peacock, J, Berrill, C and Barnard, M (Eds.) *The Print and Production Manual* Blueprint, UK (1992)
Radford, J and Richardson, D *The Management of Production* Macmillan, UK (1987)
Crosby, P B *Quality Without Tears: The Art of Hassle Free Management* McGraw Hill, USA (1986)
Guidance Notes for BS5750 Part 2 – ISO 9002 Printing Indus-

tries Quality Assurance Council Institute of Printing, Tunbridge Wells, UK (1992)

Evans, J, Anderson, R, Sweeney, D and Williams, T *Applied Production and Operation's Management* West Publishing, USA (1990)

Chapter 7 Managing printers' materials resources

Baily and Farmer *Materials Management Handbook* Gower, UK (1986)

Carter, R and Price, P *Integrated Materials Management* M & E Handbooks, Pitman, UK (1990)

Bailey, P and Farmer, D *Purchasing Principles and Techniques* Pitman, UK (1988)

Greene, J *Production and Inventory Control Handbook* McGraw Hill, UK (1987)

Chapter 8 Law and the printer

Lewis, R *Labour Law in Great Britain* Basil Blackwell, UK (1993)

Smith, J and Wood, J *Industrial Law* Butterworth, UK (1991)

Leighton, P *The Work Environment, the Law of Health, Safety and Welfare* The Industrial Society Press, UK (1991)

Oliver, M *Company Law* Pitman Books, UK (1985)

Introducing COSHH Health and Safety Executive, UK (1992)

Anson, G *Law of Contract* Oxford University Press, UK (1984)

Drake, C *Labour Law* Sweet and Maxwell, UK (1988)

Carthy, C *The New Health and Safety Handbook* Industrial Society Press, UK (1993)

Shaw, J *European Community Law* Macmillan, UK (1993)

Chapter 9 Managing change

Burns, T and Stalker, G *Management of Innovation* Tavistock, UK (1961)